# Flies for
# WESTERN
# SUPER HATCHES

# *Flies for* WESTERN SUPER HATCHES

**JIM SCHOLLMEYER**
**TED LEESON**

STACKPOLE
BOOKS

Published by
STACKPOLE BOOKS
5067 Ritter Road
Mechanicsburg, PA 17055
www.stackpolebooks.com

Printed in China

First edition

10  9  8  7  6  5  4  3  2  1

**Library of Congress Cataloging-in-Publication Data**

Schollmeyer, Jim.
  Flies for Western super hatches / Jim Schollmeyer, Ted Leeson. — 1st ed.
      p. cm.
  Includes index.
  ISBN-13: 978-0-8117-0663-6 (hardcover)
  ISBN-10: 0-8117-0663-X (hardcover)
  1. Trout fishing—West (U.S.) 2. Flies, Artificial. 3. Aquatic insects—West (U.S.) I. Leeson, Ted. II. Title.
  SH464.W4S36 2011
  688.7'9124—dc23
                                              2011016359

# Contents

Introduction    xvii

## CHAPTER 2: CADDISFLIES  119

# CHAPTER 6: OTHER IMPORTANT FOOD FORMS   273

# Thumbnail Index

## BLUE-WINGED OLIVE PATTERNS

Super Hair *Baetis*, p. 3

Bighorn Wonder Nymph, p. 4

BWO CDC Floating Nymph, p. 6

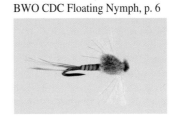

BWO Puff Fly, p. 8

*Baetis* CDC Compara-dun, p. 10

BWO Snowshoe Dun, p. 13

Parachute BWO, p. 15

BWO Hare Spinner, p. 17

## PALE MORNING DUN PATTERNS

Pheasant Tail Nymph, p. 21

Juracek Emerger, p. 23

Bat Wing Emerger, p. 25

Sparkle Dun, p. 27

CDC and Biot PMD, p. 29

PMD Hare-Wing Dun, p. 31

PMD Hare Spinner, p. 33

PMD Snowshoe Spentwing, p. 33

## SMALL WESTERN GREEN DRAKE PATTERNS

Gold-Ribbed Hare's Ear, p. 36

CDC Loop Wing Emerger, p. 38

Compara-emerger, p. 39

Little Green Drake, p. 41

Hairwing Dun, p. 43

Brown Parachute Spinner, p. 44

## TRICO PATTERNS

Trico Nymph, p. 47

CDC Female Trico Emerger, p. 48

Klinkhamer Female Trico, p. 49

Snowshoe Female Trico Compara-dun, p. 50

Parachute Female Trico Dun, p. 52

Parachute Trico Spinner, p. 53

Poly-Wing Spinner, p. 54

## WESTERN GREEN DRAKE PATTERNS

Green Drake Rubber Legs, p. 57

CDC Outrigger Emerger, p. 60

Green Drake Quigley Cripple, p. 61

Foam-Bodied Green Drake Parachute, p. 63

Green Drake Wulff, p. 66

## WESTERN MARCH BROWN PATTERNS

Pheasant Tail Nymph, p. 69

March Brown Warren Emerger, p. 70

March Brown Flymph, p. 71

March Brown Compara-dun, p. 73

Hairwing March Brown, p. 74

Soft-Hackle Spinner, p. 75

## BROWN DRAKE PATTERNS

Beadhead Strip Nymph, p. 77

Buoyant Brown Drake, p. 79

Poly-Foam Spinner, p. 83

## *CALLIBAETIS* PATTERNS

Marabou Nymph, p. 87

*Callibaetis* Emerging Dun, p. 88

Speckle-Wing Parachute, p. 90

*Callibaetis* Gathered-Hackle Spinner, p. 92

## GRAY DRAKE PATTERNS

Gray Drake Shellback, p. 95

Parachute Adams, p. 97

Clipped-Hackle Spinner, p. 99

## HEXAGENIA PATTERNS

Hex Strip Nymph, p. 101

Buoyant Hex Dun, p. 102

Buoyant Hex Spinner, p. 103

## MAHOGANY DUN PATTERNS

Pheasant Tail Nymph, p. 106

Parasol Marabou, p. 106

Mahogany Thorax Dun, p. 108

Gathered-Hackle Mahogany Spinner, p. 111

## PALE EVENING DUN PATTERNS

Thin Skin Mayfly Nymph, p. 113

Rose's Sulphur Emerger, p. 115

Parachute Light Cahill, p. 116

Gathered-Hackle PED Spinner, p. 118

## GREEN SEDGE PATTERNS

Krystal Flash Green Rock Worm, p. 121

Stretch Lace Rock Worm, p. 122

Latex Caddis Pupa, p. 124

Deep Sparkle Pupa, p. 126

Partridge and Green, p. 128

Parachute Caddis Emerger, p. 130

Deer Hair Caddis, p. 131

Snowshoe Canoe Fly, p. 133

Diving Caddis, p. 134

CDC and Wire Caddis, p. 136

## MOTHER'S DAY CADDIS AND BLACK CADDIS PATTERNS

Beadhead Prince Nymph, p. 141

Stretch Lace Caddis Pupa, p. 143

CDC Caddis Emerger, p. 146

CDC and Elk, p. 147

Black Snowshoe Caddis, p. 149

Calm-Water Caddis, p. 150

Mother's Day Caddis, p. 152

CDC Spent Caddis, p. 153

## SMALL BLACK CADDIS PATTERNS

Krystal Flash Pupa, p. 157

Super Pupa, p. 159

Super Caddis Emerger, p. 160

Black Snowshoe Emergent Caddis, p. 162

Small Black Snowshoe Caddis, p. 164

Antron Calm-Water Caddis, p. 164

CDC Small Black Caddis, p. 165

Krystal Flash Soft-Hackle, p. 166

## SPOTTED SEDGE PATTERNS

Wired Caddis Larva, p. 170

Stretch Lace Larva, p. 171

Shellback Sedge Pupa, p. 173

Beadhead Caddis Pupa, p. 175

Partridge Caddis Emerger, p. 176

X-Caddis, p. 178

Deer Hair Caddis, p. 179

EZ Caddis, p. 180

Canoe Fly, p. 181

Hare's Ear Flymph, p. 182

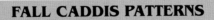

## FALL CADDIS PATTERNS

Big Orange Pupa, p. 186

Dark Caddis, p. 187

## GOLDEN STONEFLY PATTERNS

Brown Shellback Stonefly, p. 191    Basic Brown Stone, p. 193    Yellow Stimulator, p. 195    Dirty-Yellow Chugger, p. 197

## SALMONFLY PATTERNS

Shellback Salmonfly, p. 202    General Stone, p. 204    Henry's Fork Salmonfly, p. 206    Rogue Foam Giant Stone, p. 209

## *SKWALA* PATTERNS

BH Squirrel Nymph, p. 215    Foam *Skwala*, p. 216    *Skwala* Madam X, p. 218

## SUMMER STONEFLY PATTERNS

Stretch Lace Nymph, p. 223    Mojo, p. 225    Dirty-Tan Chugger, p. 227

## YELLOW SALLY PATTERNS

Bird's Nest, p. 229    Partridge and Yellow, p. 231    Elk Hair Caddis, p. 233    Henry's Fork Yellow Sally, p. 235

## MIDGE PATTERNS

Brassie, p. 239

Krystal Flash Midge, p. 240

Jujubee Midge, p. 241

Beadhead Thread Midge Pupa, p. 243

Puff Midge, p. 245

Midge Emerger, p. 246

Parachute Midge, p. 247

Snowshoe Midge Cluster, p. 248

Spent Midge, p. 248

## TERRESTRIAL PATTERNS

Thread Ant, p. 253

Fur Ant, p. 254

Delta Ant, p. 256

Parachute Ant, p. 257

Foam Beetle, p. 260

Crowe Beetle, p. 262

Parachute Hopper, p. 265

Tan Chugger, p. 268

LaFontaine Spruce Moth, p. 270

Snowshoe Moth, p. 271

## AQUATIC WORM PATTERNS

San Juan Worm, p. 274

Plastic Worm, p. 275

## BAITFISH, SMALL TROUT, AND MINNOW PATTERNS

Clouser Minnow, p. 278

Zonker, p. 280

## SCULPIN PATTERNS

Marabou Muddler, p. 284

Bunny Sculpin, p. 287

## FISH EGG PATTERNS

Hot Glue Egg, p. 291

Glo Bug, p. 293

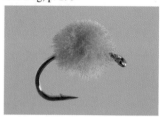

## LEECH PATTERNS

Beadhead Woolly Bugger, p. 296

Conehead Bunny Leech, p. 298

## SCUD PATTERNS

Sparkle Scud, p. 301

Brush Scud, p. 303

## SOW BUG PATTERNS

Sow Bug, p. 305

Ray Charles, p. 307

# Introduction

The contemporary fly tier lives in a world of unprecedented riches. The rapid and widespread exchange of ideas now possible in the information age has accelerated tying developments of all sorts, most particularly in fly designs. The universe of fly patterns is expanding at an astonishing rate, and tiers everywhere have benefited from these new ideas. But this growth has also magnified a dilemma that has always lurked in the background of both fly tying and fishing: so many patterns, so little time. Tiers and anglers everywhere eventually confront the same question: given the wealth of options, which patterns are worth investing the time and effort, both at the vise and on the water?

Our aim in writing this book has been to offer an answer to this question by providing a practical, straightforward reference to imitating the important food forms available to trout in the rivers and streams of the West—the region extending from the Rocky Mountains to the Pacific coast and the one that arguably contains the most popular and productive trout water in the United States.

This kind of undertaking naturally centers around aquatic insects, and most of this book focuses on what are sometimes called the "super hatches," those insects—mayflies, caddisflies, stoneflies, midges, and terrestrials—of greatest significance to Western anglers. These are the hatches that regional fly shops keep track of, local anglers mark on their calendars, and visiting anglers come to fish. Not all of these insects, however, are equally important to all trout or all trout fishermen in the West; some super hatches, you might say, are more "super" than others. To differentiate between major and secondary hatches in this book, we have divided each insect category into two groups. The "major hatches" include those insects that are so widespread and abundant that virtually no angler can afford to venture astream without patterns to represent at least some of them; Blue-Winged Olives and midges are typical examples. The "locally important hatches" represent insects that may be widely, but not uniformly, distributed throughout the region. These insects do not hatch in fishable numbers everywhere in the West but are important in particular geographic locales where the populations are dense; *Hexagenia* mayflies and fall caddis are good examples. Making this distinction between major and secondary hatches admittedly involves judgment calls, but between the two categories, we are confident that the book represents the most consequential trout-stream food forms in the West.

Since we intended this book to be a practical tying guide rather than an extended catalog of patterns, we had to limit the number fly designs. For most of the major hatches, we provide two patterns to imitate each fishable stage of the insect's life cycle, such as the nymph, emerger, dun, and spinner phases of mayflies. In a few cases, where a specific stage of the life cycle offers only marginal fishing opportunities, we include only a single fly pattern. For the locally important hatches, we give two patterns for the stages of the life cycle that are of greatest consequence to fishermen, and one pattern for each of the other fishable phases of the life cycle.

Important as they are, however, insect emergences aren't the whole story on Western trout streams, so we have also included imitations of other important food forms, such as baitfish, scuds, sow bugs, and other prey. Though not "hatches" in the customary sense, these organisms can occur in sufficient densities that the trout exhibit "hatchlike" behavior, becoming selective in feeding and requiring judicious pattern selection.

Even with the limits we placed, this book contains nearly 150 fly patterns detailed in over 100 instructional sequences. As a group, these patterns are not intended to represent standard operating equipment for every Western fly fisher, some essential "Western fly box," but rather a reservoir from which anglers can choose a selection of flies tailored to the hatch profile of the specific waters they fish.

Our primary criterion in selecting patterns for this book was their effectiveness, and our choices represent a combined six decades of experience fishing Western streams, experimenting with fly designs, and observing and photographing natural food forms, in addition to the ongoing exchanges we've had during those years with guides, anglers, and fly tiers who live in the region. At the same time, in choosing the flies, we took into account secondary criteria, important to both anglers and tiers—the versatility of a pattern, for instance; the ease and speed of tying; and an overall selection that represents a cross section of the basic categories of fly design in contemporary tying. We would hardly claim that each pattern in this book invariably represents the "best" imitation of its type for every trout stream in the western United States; the fisheries and water types are far too diverse. But we do believe the patterns here are immensely productive ones and highly credible choices on a wide range of waters.

In the matter of patterns, a word about color is in order. Many of the insect species pictured in this book show rather wide variations in color depending on the river of origin or, in some cases, the maturity of an individual insect; some caddis, for instance, are quite brightly colored when freshly hatched but darken over the course of days. Tiers often disagree about the role that color plays in the effectiveness of a fly pattern. Some scrupulously try to match the precise shade of the living insect; others believe that getting reasonably close to the color of the natural is good enough. We fall somewhere in between, depending on the specific hatch and, to some extent, the fishing pressure on a given river; at times, a careful color match does seem to give you an advantage. The most reliable way to make such a match is to capture an insect from your fishing water and use it as a tying guide. The colors of the materials listed for any pattern in this book can be altered to better correspond to the natural you've collected.

Along these same lines, it's worth calling attention to what observant fly fishers already know—that on many trout stream insects, in both the aquatic and terrestrial phases, the top of the body is a different color from the bottom. The body color that a trout sees looking up at a floating insect is not necessarily the same as the angler sees looking down at it. For tiers who wish to consider color selection in a pattern from this trout's-eye view, we've included photographs of the undersides of many of the insects that appear in this book. These photos can also provide useful information about insect profiles for tiers who wish to modify existing fly patterns or design their own.

The instructional sequences in this book assume a working familiarity with basic fly-tying procedures—mounting the thread, affixing simple materials to the hook, dubbing, finishing the head of the fly, and a few other fundamental operations. The less elementary techniques—dressing a collar or parachute hackle, for instance—are explained step by step the first time they appear. In subsequent tying sequences that employ the same technique, the text directs the reader back to the original step-by-step explanation—an approach that gives less experienced tiers the necessary detail while streamlining the instructional presentations for more experienced ones.

In the end, choosing a limited number of imitations for each natural food form proved to be a challenge, given the enormous variety of options in fly patterns these days. Yet eventually every tier faces decisions about which patterns make the most sense to tie and fish. Making such selections was how this book began, and we hope that the choices we arrived at prove useful for other fly tiers and fishermen.

Jim Schollmeyer
Ted Leeson

# Chapter One

# *Mayflies*

## MAJOR HATCHES

### Blue-Winged Olive

These small mayflies are found in virtually all Western trout streams, with the largest populations occurring in tailwaters and spring creeks. The nymphs inhabit most types of moving water, though you most often find them in riffles, runs, and flats. The nymphs mature quickly, and consecutive generations of the insect can produce fishable emergence periods throughout the year in waters that remain ice free. While fishing is generally best in the spring and fall months, more intrepid anglers can have excellent fishing to the Blue-Winged Olive (BWO) hatch in very cold, blustery, and even snowy weather. And on many waters, these hardy little mayflies provide one of the few opportunities for dry-fly fishing during the long, lean winter months. Because Blue-Winged Olives hatch nearly year-round and are so widely distributed, and because every stage of the life cycle—nymph, emerger, dun, and spinner—is available to the trout, many anglers consider the BWO to be the best mayfly hatch in the West.

**Top:** *Blue-Winged Olive nymph: olive to dark brown.*

**Bottom:** *Blue-Winged Olive nymph (underside): olive to olive-brown abdomen; thorax from lighter shade of abdomen color to tan to amber.*

| | |
|---|---|
| **Other common names:** | *Baetis*, BWO, Little Olive, Little Blue Dun |
| **Family:** | Baetidae |
| **Genus:** | *Baetis* |
| **Emergence:** | January through December, whenever the water is ice free; midday during the cooler months and in the morning during the warmer months |
| **Spinnerfall:** | Anytime from midday to evening |
| **Body length:** | ⅛–½ inch (3–12 mm) |
| **Hook sizes:** | #16-22, with #18-20 most common |

1

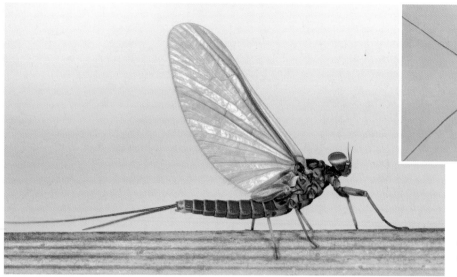

**Left:** *Male Blue-Winged Olive dun: body runs from shades of olive to olive-brown to yellow-olive to gray; light gray to dark gray wings.*

**Above:** *Male Blue-Winged Olive dun (underside): shades of olive, yellow-olive, brown, or gray, sometimes with orange highlights on thorax.*

**Left:** *Female Blue-Winged Olive spinner (shown): olive-brown to dark reddish brown. Male spinner (not shown): same shades as female, but part of the abdomen is often translucent. Clear wings on both sexes.*

**Above:** *Female Blue-Winged Olive spinner (underside): shades of olive, or tan to brown, sometimes with orange highlights on thorax.*

# Important Fishing Stages

**Nymph** Because Blue-Winged Olive nymphs are active swimmers and occur in abundance in many waters, they are frequently found drifting in the current. During nonhatch periods, fish a nymph pattern close to the bottom in riffles and runs. When the nymphs mature, their wing pads darken—your signal that an emergence period is arriving. The nymphs drift or swim to the surface and are available to trout at any point during the ascent. Unless there are large numbers of duns on the surface during a hatch, the trout tend to feed more on the nymphs, and a nymph imitation fished anywhere in the water column can often be more productive than a dry fly. To keep a nymph pattern close to the surface during a hatch, fish it behind an indicator or as a dropper trailing a dry fly.

**Emerger/dun** After mature nymphs reach the top of the water column, they break through the surface film, and the duns emerge. Both the duration of this emergence and the length of time the duns drift on the water before flying off vary with the weather; these intervals are shorter during the warmer months and longer during the cooler months. Emerger and dun patterns can be productive during hatches at any time of the year, but they tend to be most effective during the cooler weather of spring and fall, when fewer other insects are hatching. Trout often feed selectively on either emergers or adults, so it's worth studying the riseforms to determine whether the trout are taking insects on the surface or just below the film. During a sparse hatch, duns and emergers drift with the current and collect in eddies, tailouts, and seams; the accumulation of insects makes these localized areas highly productive for anglers when fewer flies are hatching.

**Spinner** Generally speaking, male spinners fall to the surface only if the mating flight has occurred over the water. Females lay their eggs on the surface or crawl underwater to deposit them on submerged objects, so trout find Blue-Winged Olive spinners both on and beneath the surface. These small mayflies, lying spent on the surface, are extremely difficult to see unless you are right next to them, and the submerged ones are impossible to detect without an insect seine. Trout feeding on or just beneath the surface when no insects are visible on the water are often your best indication that the fish are taking spinners. A Blue-Winged Olive spinnerfall can occur at the same time that the mayflies are hatching, and if an emerger or dun pattern fails to interest rising fish, a spinner pattern is often the solution. Like the emergers and duns, the drifting spinners collect in seams, tailouts, and especially eddies, and you should scout these areas for rising fish.

# Blue-Winged Olive Nymph Patterns

## SUPER HAIR *BAETIS*

| | |
|---|---|
| **Hook:** | #16-18 standard nymph |
| **Thread:** | Light brown or dark brown 8/0 |
| **Tail:** | Light brown mottled hen-hackle barbs |
| **Abdomen:** | Green or brown Super Hair |
| **Wing case:** | 4–8 strands of Black Krystal Flash |
| **Thorax:** | Brown or dark brown dubbing |

Super Hair gives the abdomen on this pattern a translucent tint that blends with the color of the thread underbody. Change the color of the thread, the Super Hair, or both to match that of the natural. When preparing the wing-case material, bundle the strands of Krystal Flash, clip them even on one end, and apply a small drop of a gel-type cyanoacrylate glue, such as superglue or Zap-A-Gap; after you form the wing case and clip the excess material, the glue keeps the strands together, ready for the next fly. On hooks smaller than #18, the Super Hair overbody will make the abdomen disproportionately plump. For these small hooks, omit the overbody in steps 2–3 and form a slender, tapered abdomen using only the tying thread. After the fly is finished, coat the thread abdomen with Sally Hansen Hard as Nails clear nail polish to protect the thread from fraying or breaking and to add a little sheen to the body. Use 2 or 3 coats, allowing them to dry between applications. Either body material will create a durable, slim-profile fly that easily penetrates the surface film without added weight. Fish the pattern just beneath the surface, as a dropper behind a dry fly, or fish it deep by trailing it behind a larger nymph or adding split shot to the leader.

**1.** Mount the thread behind the hook eye, and wrap a thread foundation to the rear of the shank. Align and mount 3 to 8 hen-hackle barbs to form tails one hook gap in length. Clip the excess, and return the thread to the rearmost thread wrap.

**2.** Mount one Super Hair strand atop the hook shank. Flatten the tying thread by spinning the bobbin counterclockwise (when viewed looking down at the bobbin barrel). Then form a thin, smooth, slightly tapered abdomen that extends forward just beyond the midpoint of the hook shank. Position the thread in front of the abdomen.

**3.** Using close, tight wraps, advance the Super Hair strand forward over the abdomen to the hanging thread. Tie off the strand.

**4.** Trim the excess Super Hair. Mount 4 to 8 strands of Krystal Flash atop the hook shank directly in front of the abdomen. While keeping the strands centered atop the hook shank, wrap the thread rearward, back over the front edge of the abdomen to the midpoint of the hook shank. Trim the excess.

**5.** Dub a thorax that is slightly larger in diameter than the abdomen. Stop about 4 thread wraps' distance behind the hook eye. Position the thread in front of the thorax.

**6.** Pull the Krystal Flash strands forward over the top of the thorax to form a smooth, shell-like wing case. Secure the strands with 4 tight thread wraps.

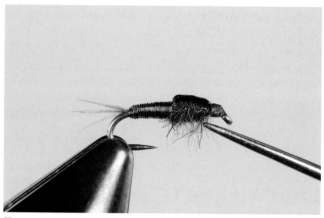

**7.** Clip the excess Krystal Flash, and finish the head. Use a dubbing needle to pick out the thorax dubbing to form the legs.

## BIGHORN WONDER NYMPH

*Originator: Brad Downey*

| | |
|---|---|
| **Hook:** | #16-20 scud or standard nymph |
| **Thread:** | Dun 8/0 |
| **Tail:** | Wood-duck flank-feather barbs |
| **Abdomen:** | Light olive-brown dubbing |
| **Wing case:** | Webby base of dun hackle feather |
| **Thorax:** | Dark olive-brown dubbing |

The Wonder Nymph is an efficient design from both the tying and fishing standpoints. It is easy to dress and requires only a few materials, one of which is the fluffy base of a hackle feather that most tiers discard as waste. And on the water, it is a highly versatile pattern. It can be fished deep, as a nymph imitation, behind an indicator and with weight on the leader; the webby base of the hackle suggests both the wing case and legs of the insect. The pattern can also be used to imitate a dun just beginning to emerge from its nymphal shuck, a sunken stillborn dun, or a sunken egg-laying or spent female spinner. During a hatch or spinnerfall, fish it in or just below the surface film by trailing it behind a more visible dry fly. To keep the fly in the film, treat it with a powder-type floatant; do not use a paste floatant, which will mat down the fluffy barbs. When used to imitate a nymph, sunken stillborn dun, or sunken spinner, it can be productively fished anywhere in the water column.

**1.** Mount the thread behind the eye, and wrap to the tailing point. Align and mount 6 to 8 wood-duck flank barbs to form tails about one hook gap in length. Trim the excess. Dub a slightly tapered abdomen to the midpoint of the hook shank. Position the thread directly in front of the abdomen.

**2.** Select a hackle feather with fluffy barbs at the base; the barbs should be about one hook gap in length. (For smaller flies, it may be necessary to trim the barbs, as shown in step 6.) Moisten your thumb and forefinger; beginning at the base of the feather, preen back a section of barbs about one hook shank in length.

**3.** Position the feather atop the shank with the tip extending over the hook eye. The concave side of the feather should face upward, and the gap you formed between the barbs in step 2 should lie directly above the hanging thread. Mount the feather atop the shank, taking 2 to 3 thread wraps in the gap, around the hackle stem only. Then secure with additional wraps toward the hook eye.

**4.** Trim the excess feather. Dub the thorax, stopping about 4 to 5 thread wraps' distance behind the hook eye. Position the thread in front of the thorax.

**5.** Fold the feather over the thorax, as shown in this top view. While holding the feather in position, use your fingers or a dubbing needle to form a gap in the feather barbs, on both sides of the feather stem, at the front of the thorax. Bind down the hackle stem exposed in this gap.

**6.** Trim the excess feather and finish the head. If the barbs are overly long, preen both sides of the wing case upward, and trim the barbs to about one hook gap in length.

# Blue-Winged Olive Emerger Patterns

## BWO CDC FLOATING NYMPH

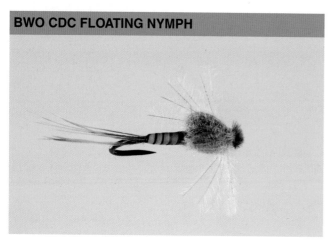

*Originator: René Harrop*

| | |
|---|---|
| **Hook:** | #16-20 standard dry-fly |
| **Thread:** | Tan 8/0 |
| **Tail:** | Dyed-dun grizzly hackle barbs |
| **Abdomen:** | Olive goose biot |
| **Loop wing:** | 2 dun CDC feathers |
| **Thorax:** | Olive-brown dubbing |
| **Legs:** | Olive CDC feather barbs |

This transitional pattern, typically fished during a hatch, imitates a nymph that has reached the surface; the nymphal skin is split, and the dun is beginning to emerge from it. The nonbuoyant biot body penetrates the film easily and hangs just beneath the surface, supported by the floating cul de canard (CDC) loop wing. The wing can be treated with a powder-type floatant to help keep it afloat, but do not use a paste floatant, since it will mat down the CDC barbs. In smooth water, the loop wing is relatively easy to see, but it can be difficult to spot on a choppy surface or in low light. Under these circumstances, fishing it as a dropper behind a more visible dry fly or using an indicator will help you present the fly accurately and detect strikes.

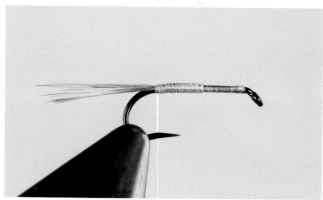

**1.** Mount the thread behind the hook eye, and wrap to the rear of the shank. Align 8 to 10 hackle barbs, and mount them atop the shank to form a tail about one hook shank in length. Trim the excess. Position the thread at the rearmost tail-mounting wrap.

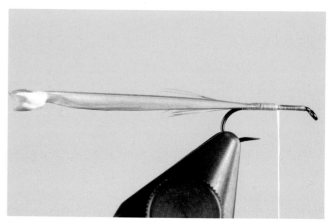

**2.** Peel a biot from the biot strip. Don't clip it off; peeling the biot leaves a notch at the base of the biot that helps orient the material for proper mounting. Mount the biot by the tip on the near side of the shank, so that the notch at the base of the biot points downward. Secure the biot tip by wrapping forward, and position the thread at the midpoint of the shank. As you wrap, spin the bobbin as needed (counterclockwise looking down at the barrel) to keep the thread flat. A smooth underbody will simplify wrapping the biot and improve the appearance of the finished fly.

**3.** Wrap the biot once around the shank. Notice the hairlike fringe that appears at the front edge of the biot.

**4.** Take another wrap of the biot, using the rear edge of the new wrap to overlap and cover the fringe raised on the first wrap. Note that a second fringe now appears at the front edge of this new wrap.

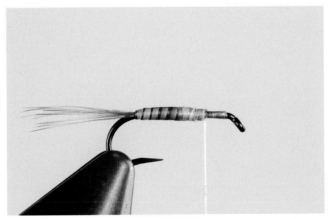

**5.** Continue wrapping the biot, using the rear edge of each new wrap to cover the fringe produced on the previous wrap. When you reach the tying thread, secure the biot, clip the excess, and bind down the butt of the material. Position the tying thread midway between the front of the abdomen and the hook eye. If you wish, you can improve the durability of the biot body by applying a thin coat of head cement.

**6.** Stack two CDC feathers together with their tips aligned and the curvatures matched. Position them atop the hook shank with the concave sides facing upward and the tips extending over the hook eye. Mount the feathers using two firm, but not tight, thread wraps.

**7.** Put slight downward tension on the bobbin with your right hand. With your left fingers, pull the feather butts rearward, sliding the feathers beneath the two mounting wraps; stop when the feather tips projecting forward beyond the thread wraps are about one shank length long.

**8.** Wrap the thread rearward to the midpoint of the hook shank, binding down the feathers as you wrap. Allowing the two feathers to splay slightly, as shown in this top view, will produce a broader, more substantial loop wing.

**9.** Trim away the feather tips. Dub the thorax to a point about 7 thread wraps' distance behind the hook eye.

**10.** Select and bundle 4 to 8 CDC barbs that are about 1½ shank lengths long. Using crisscross thread wraps, mount the CDC barbs atop the shank, directly in front of the thorax, as shown in this top view. Then apply a small amount of dubbing around the legs, stopping about 4 thread wraps' distance behind the hook eye. Dub sparsely, just enough to cover leg-mounting wraps without building excess bulk behind the hook eye.

**11.** Fold the CDC feathers over the top of the thorax so that they form a loose loop. If you have difficulty forming the loop, position a dubbing needle crosswise just above the thorax, and fold the feathers snugly over the needle to form a smooth, uniform bubble. Secure the feathers behind the hook eye with 5 tight thread wraps.

**12.** Lift the butt ends of the feathers, and take 2 or 3 tight thread wraps against the base of the feathers. Finish the head of the fly around the shank behind the eye. Trim the excess CDC to leave a small tuft over the hook eye.

**13.** Trim the legs to about ⅔ the length of the hook shank, as shown in this bottom view. Note how the CDC forms a smooth hemisphere over the top half of the fly.

**BWO PUFF FLY**

| | |
|---|---|
| **Hook:** | #16-22 standard dry-fly |
| **Thread:** | Brown 8/0 |
| **Tail:** | Dun CDC feather barbs |
| **Body:** | Fine olive-brown dubbing |
| **Wing:** | Dun CDC feather barbs |

This puff-style fly ranks among the most effective emerger/cripple designs that we've used. The fly shown here, tied with a puff wing about ¾ of a shank length in height, is reasonably visible on flat water, but in broken water, fishing the pattern behind an indicator or as a dropper behind a dry fly will help you better track the drift. When fishing very smooth, shallow water or casting to unusually finicky trout, you can trim the puff shorter to give the pattern a lower profile and more compact silhouette. The fly is moderately durable, but the CDC fibers can eventually break or become so contaminated with debris that refloating the fly is difficult. But these drawbacks are offset by the ease and speed with which the pattern can be dressed. The style is particularly well suited to smaller hook sizes that are impractical for other emerger styles. Treat the pattern with a powder-type floatant only. To keep the body slim, use a soft, fine-textured dubbing such as superfine poly or rabbit fur with the guard hairs removed.

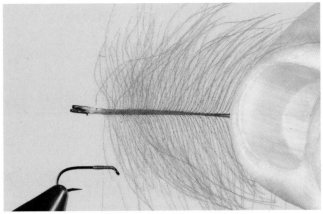

**1.** Mount the thread behind the hook eye, and form a thread base over the front half of the hook shank. Position the thread ⅓ of a shank length behind the hook eye. Select 2 to 4 CDC feathers with barbs of approximately twice the shank length. Strip away any short, scraggly barbs at the base of each feather. Stack the feathers so that the curvatures match. Hold the feathers by the tips, and preen the barbs to stand perpendicular to the stems.

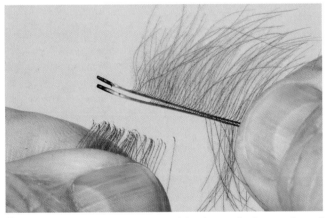

**2.** While still holding the aligned feathers, pinch a section of barbs along the lower portion of the feather; pull these barbs toward the feather butt, stripping them from the stems. Keep these barbs pinched in your fingers.

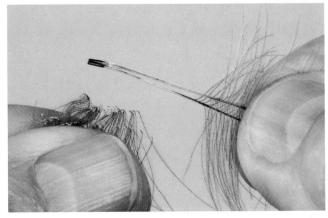

**3.** Turn the feathers over, and carefully slip another section of barbs between your thumb and forefinger, in front of the fibers you stripped previously. Pinch the barbs, and strip them from stem as explained in the previous step.

**4.** Gather the barbs into a bundle, and position the midpoint of the bundle atop the shank, above the hanging thread. Use 4 tight thread wraps to secure the bundle to the top of the shank. If more barbs are needed, repeat steps 1–3.

**5.** Pull the barbs upward, and take 3 to 5 thread wraps horizontally around the base to consolidate the barbs and post them to the vertical.

**6.** Wrap the thread to the rear of the hook shank. Mount a sparse bundle of CDC barbs, and trim them so that they extend about one shank length beyond the mounting wraps. Trim the excess.

**7.** Dub a slightly tapered body. Finish the head behind the hook eye.

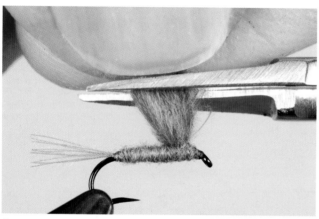

**8.** Draw the wing barbs upward, and trim them to the desired height—here, about ¾ of a shank length.

# Blue-Winged Olive Dun Patterns

### *BAETIS* CDC COMPARA-DUN

| | |
|---|---|
| **Hook:** | #16-22 standard dry-fly |
| **Thread:** | Olive 8/0 |
| **Tail:** | Dun Microfibetts |
| **Body:** | Fine olive or olive-brown dubbing |
| **Wing:** | Dun CDC feather barbs |

This small, sparse, simple fly (shown here in top view) is an excellent pattern for matching Blue-Winged Olive duns that emerge on calmer waters. It has all the virtues of the standard deer-hair Compara-dun, particularly its attractiveness to trout, but it's easier to tie and can be dressed on very small hooks, where deer hair is generally impractical. The fly lands gently on the water and is surprisingly visible for its size. Its biggest fault is that once a few trout are hooked and landed, the fly often has to be thoroughly rinsed, dried, and treated with a powder-type floatant. Even then it may not immediately regain its original buoyancy, and it frequently is best just to rinse off the fish slime, blow off the excess water, and replace the fly with a fresh one. Let the wet fly lie loose in a fly box until it has completely dried and the wing barbs are fluffed out again.

**1.** Mount the thread behind the eye, and form a thread base over the front half of the hook shank; position the thread about ⅓ of a shank length behind the eye. Strip a bundle barbs from 2 or 3 CDC feathers as shown in the instructions for the BWO Puff Fly, steps 1–3, p. 9. With the barb tips facing forward, position the center of the bundle above the hanging thread. Secure the fibers atop the hook shank with 4 tight thread wraps.

**2.** Trim the butt ends of the CDC at an angle, tapering toward the hook bend. Bind down the butt ends with the thread to form a smooth underbody. Wrap the thread to the tailing point.

**3.** Twist the thread into a tight cord by spinning the bobbin clockwise (when viewed from above). Use 3 to 6 wraps of thread to form a "bump" at the tailing point. Stack these wraps pyramid-style, wider at the base and tapering upward.

When the bump is formed, flatten the thread by spinning the bobbin counterclockwise, and position the thread directly ahead of the thread bump.

**4.** Take 1 to 3 Microfibetts, and position them atop the hook shank at a 45-degree angle, directly ahead of the thread bump. The tail should extend one shank length beyond the thread bump, as shown in this top view.

**5.** Mount the tail on the far side of the hook shank, using two thread wraps toward the hook eye.

**6.** Position a second fiber, or bundle of fibers, atop the shank to match the length and angle of the first bundle, as shown here from the top.

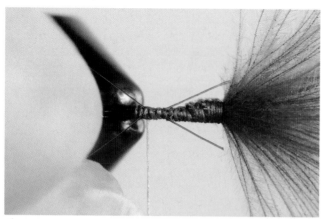

**7.** Hold this second tail in your left fingertips, and secure it by wrapping the thread rearward to the base of the thread bump, as shown in this top view.

**8.** Trim the excess tail material. Place a small amount of dubbing on the thread, and wrap a tapered body to the rear of the wing fibers.

**9.** Draw the CDC barbs upward. Bring the sparsely dubbed thread forward, and take 4 to 6 tight wraps against the front base of the barbs to raise them vertically.

**10.** Finish dubbing in front of the wing. Complete the head of the fly, and clip the thread. Draw the CDC barbs upward, and cut them to about one shank length in height.

**11.** Preen the barbs into a 180-degree arc across the top of the hook shank, as shown in this front view. Put a drop of head cement at the base of the wing to help hold the fibers in place.

## BWO SNOWSHOE DUN

| | |
|---|---|
| **Hook:** | #16-22 standard dry-fly |
| **Thread:** | Olive 8/0 |
| **Tail:** | Dun Microfibetts |
| **Body:** | Fine olive or olive-brown dubbing |
| **Wing:** | Dyed dark dun or medium dun snowshoe hare's foot hair |

In its finished appearance (shown here from the top), this fly resembles the *Baetis* CDC Compara-dun, p. 10, but a different method is used to mount and form the wing. The technique shown in the following sequence was made popular by Jim Cannon in his Bunny Dun patterns. Because it eliminates the bulky underbody associated with the standard method for winging Compara-duns, it is ideal for very small hooks, since the body can be kept slender. This winging procedure is also better suited to snowshoe hare's foot hair, which is stiffer and bulkier than CDC.

This slim-bodied, durable pattern floats extremely well even in choppy waters. After catching a trout, rinse off the slime, blot the fly dry, and if necessary apply a paste or powder floatant. Snowshoe hare's feet can differ from one another substantially, but as a general rule, the fibers found on the rear ¼ to ⅓ of the foot are too soft and fine to hold their shape well when wet, and frequently they are not sufficiently long to form the wing anyway. Use the medium-textured hair found near the center of the foot for smaller hooks and the coarser hair nearer the toe for large ones. Unlike when tying with deer hair, you need not be concerned about fibers with broken tips, as the wings are eventually trimmed to height. On overcast days or in low light, a darker wing is significantly easier to see on the water, and we always carry versions of this fly with dark dun wings for poor visibility conditions. You can also use poly yarn for this wing style, although poly yarn stains easily, taking on a green tint when it traps moss or algae from the water. Nevertheless, the material is still practical with this winging method and useful for dressing small patterns with slender bodies. To tie flies using poly yarn, begin at step 4.

**1.** Form a thread base over the front half of the hook shank, and position the thread ⅓ of a shank length behind the eye. From a hare's foot, clip a pinch of hair that is about 3 times the length of the hook shank.

**2.** Pinch the bundle of hair by the tips, and use a fine-toothed comb (a small metal comb works best) to remove the underfur. This material makes good dubbing, so you may wish to save it.

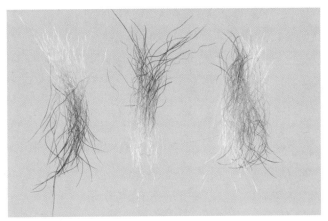

**3.** If necessary, remove additional hair so that the final bundle is ½ the desired density of the finished wing. Divide the hair into 2 equal bundles. Align the tips of one bundle with the butt ends of the other, as shown with the 2 bundles of hair on the left. Then recombine them into a single bundle, as shown on the right. Because the hair is tapered, reversing half the fibers, tip to butt, makes a bundle of more uniform density.

**4.** Center the bundle atop the shank, above the hanging thread. Secure the hair with 3 thread wraps.

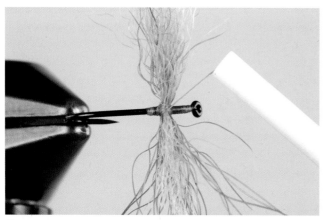

**5.** Pinch the hair tips at each end, and rotate the bundle 90 degrees so that it is positioned perpendicular to the hook shank, as shown in this top view. Take a tight thread wrap that starts behind the hair on the near side of the shank, crosses over the top of the bundle, and goes down in front of the hair on the far side of the shank. Depending on the size of the bundle, take 2 to 4 additional wraps.

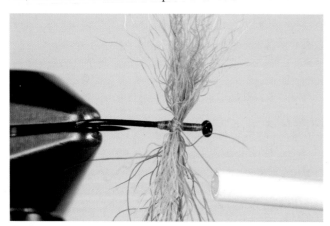

**6.** Now bring the thread from behind the hair on the far side of the shank, over the top, and down in front of the hair on the near side, as shown here from the top. Use the same number of crossover wraps as you did in the previous step.

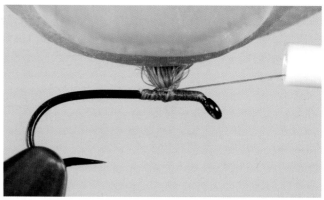

**7.** Pull all the hair upward, and take 2 snug thread wraps horizontally around the base of the hairs. These wraps will partially consolidate the hair into a single bundle again and raise it vertically. Since these wing fibers will be fanned out in the final step, don't bind them around the base as tightly as you would when forming a parachute post. You want them to remain slightly mobile.

**8.** Wrap the thread to the rear of the hook shank. Mount a split tail as shown in the procedure for the *Baetis* CDC Compara-dun, steps 3–7, pp. 11–12. Then dub a slender body.

**9.** Tie off the thread, and finish the head. Preen the hairs upward, and cut them to about one shank length in height.

**10.** Preen the hairs into a 180-degree arc over the top of the hook shank, as shown in this front view, and trim away any stray hairs. Depending on the texture of the hair used, the wings may spread in a uniform fan or be slightly divided, as they are here, with a gap or sparse area in the center. Either result is perfectly fine. Place a drop of head cement at the base of the wing to help hold the hairs in place.

# Blue-Winged Olive Spinner Patterns

## PARACHUTE BWO

| | |
|---|---|
| **Hook:** | #16-20 standard dry-fly |
| **Thread:** | Brown 8/0 |
| **Tail:** | Dun Microfibetts |
| **Body:** | Fine olive-brown or reddish brown dubbing |
| **Wing post:** | Light dun or dark dun poly yarn |
| **Hackle:** | Light dun dry-fly |

Although most anglers and tiers tend to associate parachute patterns with mayfly dun imitations, they make excellent mayfly spinner representations as well. The body sits flush against the film like that of a spinner; the light dun barbs radiating outward suggest the veined translucency of spinner wings and the crinkly shimmers of light they produce. Moreover, parachute spinners float far better than the traditional spent-wing type, are more durable, and are considerably more visible to the angler, especially in the smaller sizes. The visibility can be further improved, at least under some circum-

stances, by using an orange, pink, or red wing post, though in our experience, cautious trout can be spooked by these unnaturally bright colors. In the flat light of overcast days, the silvery glare of the water can make a light-colored wing very difficult to spot; a dark dun wing shows up much better under such conditions. Tied with an olive-brown body, this pattern also makes an effective crippled-dun imitation.

**1.** Mount the thread behind the hook eye, and wrap a thread foundation over the front half of the hook shank. Position the thread ⅓ of a shank length behind the eye. Center a 2-inch length of poly yarn that is ½ the thickness of the finished wing post atop the shank over the hanging thread. Use 4 tight thread wraps to secure the yarn to the top of the hook shank.

**2.** Spin the bobbin counterclockwise (as seen from above) to flatten the thread. Pull both ends of the yarn upward, and take 6 to 8 snug thread wraps up the post, placing each wrap directly above and butting against the previous one. Spin the bobbin as necessary to keep the thread flat. Then wrap the thread smoothly back down the post, and position it directly behind the post.

**3.** Strip about one hook shank's length of barbs from the base of a hackle feather. Position the feather with its glossy side up, across the near side of the wing post at a 45-degree angle, with the feather tip pointing rearward. The length of the bare hackle stem extending behind the wing post should be equal to the height of the thread foundation on the wing post.

**4.** Take 4 tight thread wraps in front of the wing post to secure the stem to the side of the hook shank. Trim the excess stem. To keep the hackle out of the way while tying, prop it against the front of the wing post.

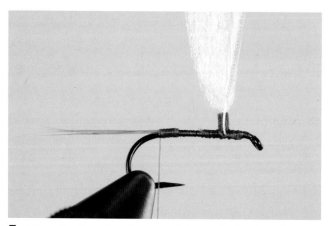

**5.** Wrap the thread to the rear of the hook shank, and mount 3 to 5 Microfibetts to form a tail about as long as the hook shank. Trim the excess, and return the thread to the rearmost tail-mounting wrap.

**6.** Dub a slightly tapered body, stopping 4 thread wraps' distance behind the hook eye.

**7.** Spiral the bare hackle stem (clockwise as seen from above) to the top of the thread foundation. Hold the feather parallel to the hook shank, with the feather barbs on a horizontal plane.

**8.** Take the first wrap of hackle at the top of the thread post, keeping the feather horizontal. Take the next wrap of hackle beneath the first one.

**9.** Take 2 to 4 more wraps of hackle around the post, placing each wrap directly beneath the previous one. As you make the last wrap of hackle, bring the feather tip over the top of the body and down the near side of the hook shank in front of the dubbing. Hold the feather in this position with your right fingers or hackle pliers.

**10.** With your left fingers, preen back the barbs to expose the feather stem at the point where it crosses the hook shank. Use your left fingers to pinch the feather—behind the exposed stem—against the hook shank. Release the feather tip, and secure the stem to the hook shank with 3 to 4 tight thread wraps.

**11.** Trim the excess feather, and finish the head. Trim the wing post so that it extends about ½ to ¾ the length of the hook shank above the topmost wrap of hackle.

### BWO HARE SPINNER

| | |
|---|---|
| **Hook:** | #16-22 standard dry-fly |
| **Thread:** | Brown 8/0 |
| **Tail:** | Light dun Microfibetts |
| **Body:** | Fine reddish brown or olive-brown dubbing |
| **Wing:** | Light dun snowshoe hare's foot hair |

Except for the color of the materials, this fly is identical to the BWO Snowshoe Dun shown on p. 13, and the tying steps to dress it are the same. It may seem a bit strange that one of our favorite Blue-Winged Olive spinner imitations is also a dun pattern, since these two stages of the insect produce such vastly different silhouettes on the water. But one of the curious mysteries of trout fishing is that at times, the trout seem to care a great deal about details of a fly pattern that seem trivial to an angler, whereas at other times, they seem indifferent to details that the angler assumes would be important to them. The use of this arc-wing pattern during a spinnerfall fits into the second category. By all conventional measures, this high-profile wing would seem better suited—and indeed is quite well suited—to a dun imitation. But while it is unconventional in appearance compared with the more sparsely dressed standard spent-wing patterns, the trout don't seem to mind at all. Certainly, the lowermost wing fibers spread out on the water do produce the crinkly, translucent impression of spinner wings against the film. But why trout frequently choose to ignore the fanlike wing above the body remains a puzzle. Yet they do—not always, but often. And we'll choose it first over spent-wing dressings, which can be difficult to see and to keep floating. The arc wing makes this spinner pattern highly visible on the surface, and the buoyancy of the snowshoe hare's foot hair makes the fly excellent on choppy or broken water.

# Pale Morning Dun

Pale morning duns (PMDs) are found in most Western trout streams. The nymphs live among the streambed rocks in the moderate flows of riffles and runs, with the largest populations generally occurring in silt-free tailwaters and spring creeks. Depending on the particular stream, a hatch cycle can be as brief as a couple of weeks or as long as a couple of months. But on any given river, PMDs are one of the more predictable mayflies, hatching during same annual interval at approximately the same time each day for about the same duration, typically an hour or two. From the angler's standpoint, and the trout's as well, PMDs are one of the better Western super hatches: the insects are widespread and emerge on a relatively consistent schedule; the hatch cycle can be long on some rivers; and in the right weather, the duns may spend an extended time drifting on the water before flying off. The body colors of the duns often vary from stream to stream, and there are frequently color differences between the males and females. The duns also vary in size, with the largest flies appearing in the early season and the smaller ones later. Because of size and color variations in these mayflies, it is often important to match emerger, dun, and spinner imitations to the sizes and approximate colors of the naturals on a stream-by-stream basis.

| | |
|---|---|
| **Other common name:** | PMD |
| **Family:** | Ephemerellidae |
| **Genus:** | *Ephemerella* |
| **Emergence:** | Mid-May to mid-October, with peak periods in June in the Pacific region and mid-July to mid-August in the Rocky Mountains. Early- and late-season hatches tend to occur from late morning to midafternoon. Summer hatches are more likely from midmorning to early afternoon. |
| **Spinnerfall:** | Can occur any time of the day, but generally during the windless hours in the early morning during warmer weather and in late afternoon during the cooler months |
| **Body length:** | ¼ inch–⅝ inch (6–12 mm) |
| **Hook sizes:** | #14-20, with #16 most common |

**Left:** *Pale Morning Dun nymph: light brown to dark brown, with reddish or olive tones.*

**Above:** *Pale Morning Dun nymph (underside): tan or shades of light brown to dark brown, with reddish or olive tones.*

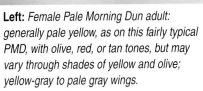

**Left:** *Female Pale Morning Dun adult: generally pale yellow, as on this fairly typical PMD, with olive, red, or tan tones, but may vary through shades of yellow and olive; yellow-gray to pale gray wings.*

**Above:** *Female Pale Morning Dun adult (underside): usually a lighter shade of the top color, in this case pale yellow, sometimes with reddish highlights.*

**Left:** *Female Pale Morning Dun: some PMDs, such as the one shown here, have a distinctly olive cast to the body.*

**Above:** *Female Pale Morning Dun (underside): the underside of the olive-bodied PMD is the same shade as the top color.*

**Left:** *Male Pale Morning Dun: generally brown as shown here, but can vary through shades of yellow and brown.*

**Above:** *Male Pale Morning Dun (underside): usually the same color as, or a lighter shade of, the top color, as on this brown-bodied adult.*

**Above:** *Female Pale Morning Dun spinner: brown to reddish brown to pale olive-yellow, with clear wings.*

**Top right:** *Female Pale Morning Dun spinner (underside): usually a lighter shade of the top color.*

**Right:** *Male Pale Morning Dun spinner: brown to reddish brown to darker yellow-brown, with clear wings; the underside is generally a lighter shade of the top color.*

# Important Fishing Stages

**Nymph** These nymphs are seldom found drifting in the current until just before emergence. Somewhat feeble swimmers, they can drift long distances downcurrent before reaching the surface, during which time they are highly vulnerable to the trout. As a consequence, it can be very productive, in the hour or two before a PMD emergence, to fish a nymph pattern close to the bottom in riffles and runs, and in the water downstream of these areas. If you see duns hatching but few or no visible rises, the trout are probably feeding on nymphs close to the streambed; fish a nymph pattern drag-free and deep, and at the end of the drift, let the fly swing to the surface. If you see subsurface boils, fish a nymph pattern shallow, below an indicator or behind a dry fly.

**Emerger/dun** Duns may emerge from the nymphal shuck a few inches beneath the surface, on the surface film, or anywhere in between. On heavily fished streams or during heavy hatches, trout often key on the emergers rather than the upright duns, so it is worth observing the riseforms. A surface-feeding trout generally leaves a small bubble at the center of the riseform. If you see such a bubble but saw no dun plucked from the surface, fish a floating emerger pattern. If the rise produces no bubble, the fish is feeding just subsurface, and under these circumstances, a sunken emerger pattern fished behind an indicator or dry fly is the best bet. Cripples

and stillborns are common during PMD hatches and quite numerous when the hatch is heavy; a floating or sunken emerger can be used to imitate them.

Once the duns reach the surface, their wings must dry and stiffen before the insects can fly off. How much time this takes, and consequently how far the flies drift on the surface in the meantime, varies with the weather; warmer or drier days shorten the time, whereas cooler or rainy weather lengthens it. These duns drift serenely on the surface, and a dun pattern must be presented with a drag-free drift. Trout can become very particular during this hatch, and matching a dun pattern to the size and approximate color of the natural can greatly improve your chances of taking selective fish.

**Spinner** Mating flights normally occur over riffles. After mating, the spent males often fall onto the water. The females fly to streamside vegetation, extrude their eggs, and return to the riffles, where they dip to the surface and release their eggs, eventually falling spent to the surface. A PMD spinnerfall can be a bit puzzling at times. Trout frequently take up feeding stations downstream of the area where mating and egg-laying flights occur, so you may find plenty of mayflies but no rises. Conversely, you may observe fish feeding but see no insects, since the spinners lying flush on the surface are quite difficult to spot. In fact, seeing trout rising to invisible insects is sometimes the best clue to spinner-feeding fish. Matching the size of the spinner is more important than matching the color, but at times trout do key on the color. Normally the males, which tend to be a shade of rusty brown, are the first spinners on the water; they're followed by the females, which tend to be a lighter color. So if a rusty brown spinner pattern of the right size brings refusals, switching to a pale yellow-olive spinner can make a difference. In a pinch, a dun pattern of the appropriate color will sometimes do the trick.

# Pale Morning Dun Nymph Patterns

## PHEASANT TAIL NYMPH

*Originator: Frank Sawyer/Al Troth*

| | |
|---|---|
| **Hook:** | #14-16 1X–2XL nymph |
| **Weight:** | Lead or nontoxic wire |
| **Thread:** | Brown 8/0 |
| **Tail:** | Pheasant tail-feather barbs |
| **Rib:** | Fine copper wire |
| **Abdomen:** | Pheasant tail-feather barbs |
| **Wing case:** | Pheasant tail-feather barbs |
| **Thorax:** | Peacock herl |
| **Legs:** | Pheasant tail-feather barbs |

Although Frank Sawyer designed the original Pheasant Tail Nymph, the version shown here is generally credited to Al Troth. This slim-bodied pattern is an excellent imitation of a mature PMD nymph with darkened wing pads. Before the onset of the hatch, fish this fly dead-drift, close to the streambed. During a hatch, when the nymphs are ascending to the surface, you can fish the fly anywhere in the water column. This pattern can also be tied in an unweighted version for fishing just beneath the surface.

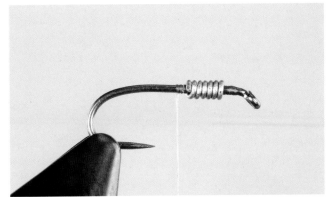

**1.** Mount the thread behind the hook eye, and wrap a thread foundation to the midpoint of the shank. For weighting the fly, use lead or nontoxic wire with a diameter no larger than that of the hook shank; thicker wire is difficult to secure firmly. Cut a 2-inch length of wire. Beginning at about the midpoint of the shank, wrap the wire tightly forward to a point about 6 or 8 thread wraps' distance behind the hook eye. Clip the excess.

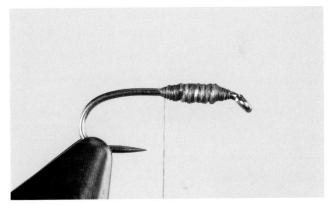

**2.** Secure the wire by building a tapered ramp of thread at each end of the wire wraps. Then wrap the thread back and forth over the wire a few times to secure it. Position the thread at the rear of the wire underbody.

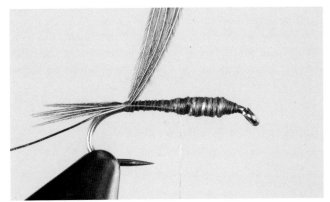

**3.** Mount a 4-inch length of copper wire at the midpoint of the shank. Wrap the thread rearward to the tail-mounting point, binding the wire to the near side of the hook shank. Align the tips of 4 to 6 pheasant tail barbs, and mount them atop the shank with 2 tight thread wraps to form tails one hook gap in length. Lift the butt ends of the barbs, and advance the thread to the midpoint of the shank.

**4.** Lift the pheasant tail barbs vertically, and try to smooth them into a flat band or ribbon; wrapping the barbs as a band, rather than a tight bundle, will keep the abdomen slim. Wrap the barbs forward to the tying thread, and secure them to the shank.

**5.** Trim the excess feather barbs. Counterwrap the copper wire forward; that is, wrap it counterclockwise when viewed from the hook eye. Wrapping in this direction helps prevent the pheasant tail fibers from unraveling if they break. Wrap the wire in an open spiral with 3 to 5 turns; tie off the wire at the front of the abdomen, and clip the excess. Align the tips of 8 to 12 pheasant tail barbs, and size them against the hook shank, as shown, to establish the mounting point. The barbs should project beyond your fingertips about one shank length's distance.

**6.** Transfer the barbs to your left hand so that your left thumb and forefinger are pinching the bundle at the mounting point. Mount the bundle, tips rearward, atop the hook at the midpoint of the shank. Position the tying thread at the rearmost mounting wrap.

**7.** Trim the excess feather barbs. Align the tips of two peacock herls and a 6-inch length of scrap tying thread, and trim away the uppermost ½ inch of the bundle. Secure the herl and tying thread atop the shank at the midpoint, and trim the excess. Position the thread 4 to 6 thread wraps' distance behind the hook eye.

**8.** Draw the herls and scrap thread downward as a bundle, and gently twist the bundle 4 or 5 times; don't be too aggressive or the fragile herls will break.

**9.** Wrap the bundle forward to the tying thread, pausing to retwist it as necessary to maintain a chenille-like strand. Finish the last wrap with the herl strand held vertically above the shank. Secure the herl with 2 or 3 tight wraps of thread.

**10.** Clip the excess herl. Gather the wing-case barbs, and draw them smoothly forward over the top of the herl thorax. Secure them with 3 thread wraps; do not clip the excess.

**11.** To form the legs, grasp 3 or 4 barbs and fold them rearward against the far side of the hook shank; take a tight thread wrap over the base of the fold to hold them in this position. Repeat with 3 or 4 barbs on the near side of the shank. Take 2 or 3 additional thread wraps over the base of the legs to hold them securely, as shown in this top view. Trim any excess barbs over the hook eye, and finish the head.

## JURACEK EMERGER

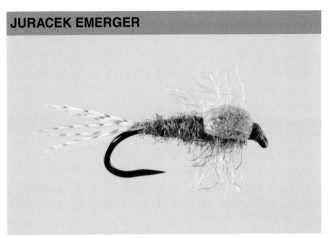

*Originator: John Juracek and Craig Mathews*

| | |
|---|---|
| **Hook:** | #14-18 standard dry-fly |
| **Thread:** | Brown 8/0 |
| **Tail:** | Mallard flank-feather barbs |
| **Abdomen:** | Dark brown sparkle dubbing |
| **Wing case:** | Dark gray foam |
| **Thorax:** | Brown sparkle or Hare-Tron dubbing |

The emergence of an aquatic insect is a process of transformation, but an emerger pattern can represent only a single point in that process. Most conventional emergers imitate a point near the conclusion of the transformation, as the adult insect is exiting the nymphal shuck and climbing out onto the surface film. This Juracek–Mathews pattern represents a point closer to the beginning of the process, so the fly looks and is fished like a nymph. Since PMDs can begin their emergence a few inches beneath the surface, the effectiveness of this fly could be due in part to the gray wing case, which suggests the gray wings of the adult just beginning to protrude through the nymphal shuck. But there's no question that the light wire hook and buoyant foam wing case cause this pattern to sink very slowly through the uppermost zone of the water column, where trout feed on nymphs during a hatch. This is a productive pattern whenever trout are taking insects within a foot or so of the surface during a PMD emergence. Fish it dead-drift on a leader greased to within a foot of the fly or as a dropper behind a more visible dry fly. This simple pattern is quite easy to tie, and almost any dubbing can be used. We prefer a dubbing that is finely textured, so that it can be twisted tight for a slim abdomen, but which also includes some stiffer fibers to pick out for legs.

**1.** Mount the tying thread behind the hook eye, and wrap rearward to the tailing point, forming a thread foundation. Align 8 to 10 barbs of a mallard flank feather, and mount them to form a tail one hook shank in length. Clip the excess.

**2.** Dub the tying thread sparsely, and form a slim, slightly tapered abdomen to the midpoint of the hook shank.

**3.** Cut a strip of foam about one hook gap in width. Secure it directly in front of the abdomen, beginning with firm but not tight wraps, and gradually increasing the thread pressure as you advance the thread forward. Clip the excess foam. Then wrap the thread tightly back to the base of the wing case.

**4.** Dub a thorax about twice the diameter of the abdomen, stopping about 6 thread wraps' distance behind the eye. Position the tying thread at the front of the thorax. Use a dubbing needle to pick out fibers on the sides of the thorax to suggest legs.

**5.** Trim the fibers picked out for legs to about one hook gap in length. Fold the foam strip forward over the top of the thorax, and secure it tightly in front of the thorax.

**6.** To avoid bulk at the head of the fly, stretch the foam tag, and trim it close to the thread wraps. Bind down the excess foam, and finish the head.

# Pale Morning Dun Emerger Patterns

## BAT WING EMERGER

*Originator: Tracy Peterson*

| | |
|---|---|
| **Hook:** | #14 fine-wire scud |
| **Bead:** | 2.3 mm gun-metal-blue glass bead |
| **Thread:** | Rusty dun 8/0 |
| **Tail:** | Dun hen-feather barbs |
| **Abdomen:** | Yellow-brown turkey biot |
| **Wing:** | Fluffy butt end of dun hen hackle |
| **Thorax:** | Yellow-olive dubbing |

This pattern makes use of a material most fly tiers throw away—the fluffy base of a hackle feather—to imitate the wings on a PMD that is emerging below the surface. The fly, however, can also represent an insect that has become stuck in the nymphal shuck and drowned before reaching the surface, or it can imitate a drowned dun. The small glass bead at the head gives the fly just enough weight to penetrate and sink beneath the surface film, and the soft, fluffy wing barbs have excellent mobility underwater. To imitate duns emerging just below the surface, fish this pattern shallow, below an indicator or as a dropper behind a dry fly. For a short time after a PMD hatch has ended, you can fish this fly at any depth, particularly along current seams and in eddies, where sunken naturals collect. You can also tie this fly in size 18, using a 2 mm bead.

**1.** Slide the bead over the hook, and position it at the eye. Mount the thread behind the bead, and wrap a thread foundation to the tailing point, about halfway around the bend. Align 4 to 6 hen-hackle barbs, and mount them atop the shank to form a tail about one shank length long. You may find mounting the tails easier if you rotate the hook forward to put the tail-mounting point upward, as shown here. Position the thread at the rearmost tail-mounting wrap.

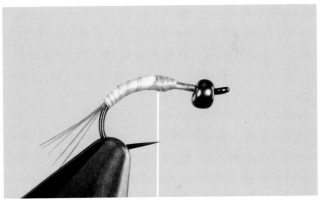

**2.** Peel a biot from a biot strip, and mount it as shown in the instructions for the BWO CDC Floating Nymph, step 2, p. 6. Rotate the hook back to the normal tying position, and form a smooth underbody to the midpoint of the shank. Form a biot body as shown in steps 3–5, pp. 6–7.

**3.** Select a hen-hackle feather, preferably one with barbs about one hook gap in length. If you can't find a feather of the right size, use one with barbs that are slightly longer; they can be shortened, as shown in the next step. Cut off the fluffy part of the base, as shown.

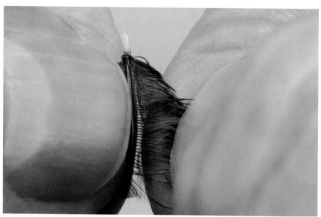

**4.** If necessary, shorten or even out the length of the fluffy barbs by pinching them off with your thumbnail to one hook gap in length.

**5.** Mount the feather by the clipped end atop the shank directly in front of the abdomen. The convex side of the feather should face upward. Dub a thorax about twice the thickness of the abdomen, leaving a space about 5 thread wraps wide between the front of the thorax and the bead.

**6.** Push the bead against the front of the abdomen. Lift the thread over the bead, and take 3 thread wraps against the front of the bead.

**7.** Fold the feather section forward over the bead. Draw the barbs behind the tie-off point rearward to form a gap and expose a short section of bare quill just above the tying thread. You may find this step easier if you dampen the feather barbs slightly. Secure the feather by binding down the bare quill exposed in the gap. Clip the excess, and finish the head.

**8.** Here's a front view of the finished fly.

## SPARKLE DUN

*Originator: Craig Mathews*

**Hook:**      #14-16 standard dry-fly
**Thread:**    Olive 8/0
**Tail:**       Brown Z-Lon
**Body:**     Yellow-olive dubbing
**Wing:**     Deer hair

This variation of the well-known Compara-dun has become widely used throughout the West. The pattern imitates an emerged dun in the final stage of transition, with the nymphal shuck, represented by the Z-Lon fibers, still attached at the rear of the adult. This pattern floats well, and the wing is highly visible to the angler. Treat the entire fly with floatant. The best deer hair for the wings has a fairly finely-textured, uniform-diameter shaft that tapers rather abruptly at the tip. Avoid hair with long, gradually tapering, threadlike tips; it lacks buoyancy and produces a poor wing silhouette. Similarly, avoid thick-diameter "spinning hair," which flares too much to make a nicely consolidated wing. Some types of deer hair labeled Compara-dun Hair, particularly those sold by Nature's Spirit and Hareline, are excellent for this pattern.

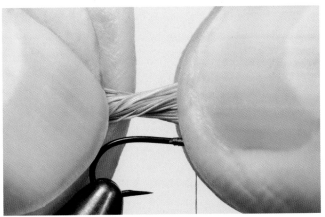

**1.** Mount the tying thread behind the eye; wrap a thread foundation to the midpoint of the shank, and then return the thread to a point ⅓ of a shank length behind the eye.

Clip a bundle of deer hair. The proper amount of hair depends on the hook size, your preferences, and to some extent the hair itself. As a starting point, try this: pinch the ends of the bundle in your right and left fingertips, then twist

the ends of the hair a quarter turn each in opposite directions. The thickness of the bundle should be approximately ½ the hook gap. If the finished wing looks too sparse or too dense, adjust the quantity of hair used on subsequent flies. Tiers have a tendency to overdress this fly by using too much hair, which makes the wing difficult to secure, produces a bulky under-body, and gives a somewhat unruly wing silhouette. We find it better to err on the side of too little hair than too much.

**2.** Pinch the hair bundle by the tips. Use your fingers or a fine-toothed comb, as shown here, to remove the soft, crinkly underfur.

**3.** Insert the bundle of hair, tips first, into a hair stacker. Hold your index finger over the stacker opening. Hold the stacker vertically, and tap the bottom of the stacker sharply against a tabletop 4 or 5 times. Coarse or crinkly hair or large bundles may take more taps. Now tilt the stacker at a 45-degree angle, as shown here, and tap a few more times; angling the stacker like this helps consolidate the hair inside.

**4.** Hold the stacker horizontally, and remove the barrel to expose the hair tips. If the hair is to be mounted with the tips over the hook eye, as on this pattern, grasp the tips in your right fingers, and pull the hair from the stacker. (For patterns on which the hair tips are mounted toward the hook bend, withdraw the hair with your left fingers. Removing the hair with the correct fingers minimizes the need to transfer it from hand to hand and helps keep the tips aligned.)

**5.** Grasp the butts of the hair in your left fingers. Position the bundle atop the shank so that the hair tips extend beyond the hanging thread a distance of one shank length; this will form wings of the proper height.

**6.** Take 2 firm, but not tight, wraps over the hair bundle. Pinch the bundle tightly to keep it atop the shank, and pull downward on the tying thread to cinch the hair against the top of the shank. Use your left index finger as a backstop to prevent the thread tension from pulling the hair to the far side of the shank.

**7.** Take 4 to 5 more tight thread wraps toward the hook bend, sliding your left fingers rearward as you go to keep the hair atop the shank. Clip the butts at an angle, as shown, and bind down with thread. Return the thread to the base of the hair bundle.

**8.** Draw the rearmost ⅓ of the hair bundle rearward. Take one very tight thread wrap in front of it.

**9.** Draw the middle ⅓ of the hair rearward, and take one tight wrap as you did in the previous step. Then draw the entire bundle of hair rearward. Bring the thread to the front of the bundle, and build a tight ramp of the thread that abuts the base of the hair and tapers toward the hook eye; these wraps will post the wing hair to the vertical.

**10.** Use your fingers to preen the hair outward to form an arc of 180 degrees, as shown in this front view.

**11.** Wrap the thread to the tailing point, and mount a sparse bundle of Z-Lon to form a shuck one hook gap in length. Clip the excess Z-Lon, and return the thread to the rearmost shuck-mounting wrap.

**12.** Apply dubbing sparsely to the thread to form a slim body. When you reach the base of the wing, take a crisscross wrap of dubbing underneath the shank to cover the wing-mounting wraps, as shown in this underside view. Dub the head, and finish the fly.

# Pale Morning Dun Dun Patterns

## CDC AND BIOT PMD

| | |
|---|---|
| **Hook:** | #14-16 standard dry-fly |
| **Thread:** | Pale-yellow 8/0 |
| **Tail:** | Dun Microfibetts |
| **Abdomen:** | Yellow-olive biot |
| **Thorax:** | Yellow-olive dubbing |
| **Wing:** | 2 medium dun CDC feathers |
| **Legs:** | Butts of CDC wing feathers |

This pattern combines a slim biot abdomen with a light-weight but full-profile CDC wing and outrigger-type legs. It's not particularly well suited to heavier water, but the feathery wing, which lands quietly on the water, and the trim body make this a highly effective pattern on smooth, glassy water, where PMD hatches can be their most challenging. It's worth carrying multiples of this pattern, since the CDC can become sufficiently contaminated with algae or fish slime that it no longer floats the fly. At that point, it's best to rinse the wing in clear water, set it aside to dry, and tie on a fresh fly. Use a powder-type floatant only.

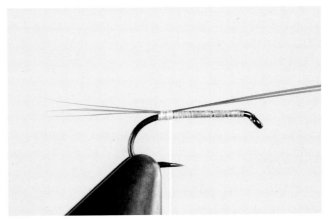

**1.** Mount the tying thread behind the hook eye, and wrap rearward to the tailing point, forming a smooth underbody. Align the tips of 4 to 6 Microfibetts, and mount them to form a tail one hook shank in length. Position the thread at the rear-most tailing wrap. Do not clip the tail butts.

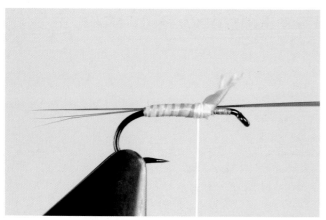

**2.** Mount a biot as shown in the instructions for the BWO CDC Floating Nymph, step 2, p. 6. Wrap the thread forward, binding down the biot tip and tail butts, forming a smooth underbody to the midpoint of the shank. Wrap the biot as explained in steps 3–5, pp. 6–7. Tie off the biot at the midpoint of the shank.

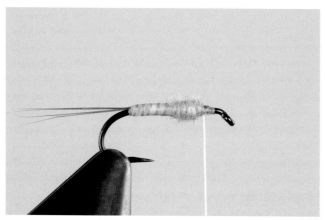

**3.** Clip the excess biot and the tail butts. With a thinly dubbed thread, form a thorax from the front of the abdomen halfway to the hook eye. Position the thread at the front of the thorax.

**4.** Stack the two CDC feathers with the convex sides together and the tips aligned. Preen the CDC barbs toward the feather tip. Position the feathers atop the shank at the front of the thorax. The broad side of the feathers (not the edge) should face you, and the feather tips should extend 2 shank lengths beyond the end of the body.

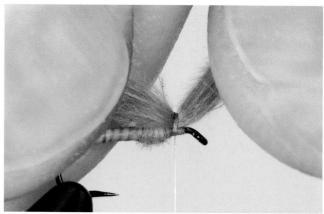

**5.** Pinch the feather tips in your left fingers. Take 2 firm, but not tight, turns of thread directly ahead of the thorax to hold the CDC feathers atop the shank. Do not clip the feather butts. While holding the feather tips with your left fingers, use your right fingers to pull the feather butts forward and slightly upward, sliding the feathers under the thread wraps.

**6.** When the feather tips are aligned with the rear of the body, secure the feathers by wrapping forward over the butts, stopping 2 thread wraps' distance behind the hook eye. Do not clip the feather butts.

**7.** Position the thread directly in front of the wings. Fold the feather butts rearward, and pinch them against the sides of the body. Take 3 tight thread wraps over the folds, as shown in this top view, to slant the feather butts rearward at an angle of about 45 degrees.

**8.** Position the thread behind the hook eye, and finish the head of the fly. Clip the feather butts to make outrigger-type legs ½ the hook shank in length, as shown in this bottom view.

## PMD HARE-WING DUN

| | |
|---|---|
| **Hook:** | #14-18 standard dry-fly |
| **Thread:** | Pale-yellow 8/0 |
| **Tail:** | Dun Microfibetts |
| **Body:** | Fine yellow-tan or yellow-olive dubbing |
| **Hackle:** | Dun dry-fly |
| **Wing:** | Medium dun snowshoe hare's foot hair |

This pattern is a variation of René Harrop's Hairwing Dun. Using snowshoe hare's foot hair offers a number of advantages over the deer hair used in the original dressing. Snowshoe hare gives a better color match for PMD wings, offers greater visibility to the angler, and is more durable. We think the material also offers better flotation on smaller hooks, where a short deer-hair wing would consist mainly of the less buoyant hair tips. In fact, even in smaller sizes, this pattern floats well enough to support a nymph or sunken emerger dropper for fishing two stages of the hatch simultaneously, and it's also suited to rougher or broken water. The fly can be treated with any type of floatant.

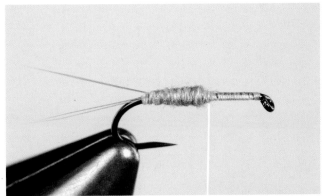

**1.** Mount the thread behind the hook eye, and wrap rearward, forming a thread foundation to the tailing point. Mount 2 to 4 Microfibetts on each side of the hook to form a split tail as shown in the instructions for the *Baetis* CDC Compara-dun, steps 3–7, pp. 11–12. Clip the excess tail material, and position the thread at the rearmost tailing wrap. Use a thinly dubbed thread to form a slightly tapered abdomen to the midpoint of the shank. Position the thread at the front of the abdomen.

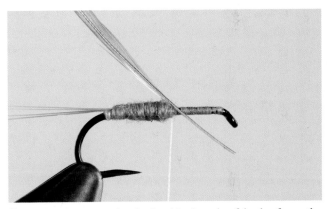

**2.** Strip about one hook shank's length of barbs from the base of a hackle feather. Position the feather edgewise as shown, with the glossy side facing down. The bare hackle stem crosses the near side of the hook shank at a 45-degree angle, directly in front of the abdomen.

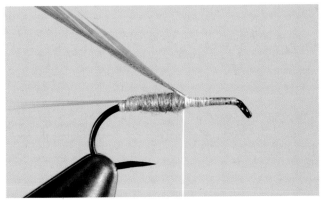

**3.** Wrap the thread forward, binding the bare hackle stem to the near side of the shank. Clip the excess hackle stem, and return the thread to the rearmost mounting wrap.

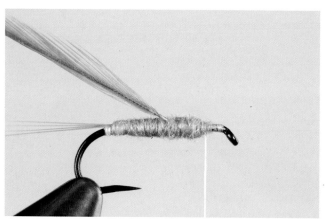

**4.** Dub a thorax that continues the body taper to a point 5 to 6 thread wraps' distance behind the hook eye.

**5.** Grasp the tip of the hackle feather in your fingers or a pair of hackle pliers, and wrap forward 3 to 5 turns in an open spiral. As you wrap, keep the plane of the hackle feather vertical and the glossy side facing the hook bend. When you reach the front of the thorax, tie off the hackle as shown in the instructions for the Little Green Drake, steps 9–10, pp. 42–43.

**6.** Prepare a bundle of snowshoe hare's foot hair as described for the BWO Snowshoe Dun, steps 1–2, p. 13. Center the bundle atop the shank, tips pointed rearward, at the front of the hackle. Secure it with 4 to 6 very tight thread wraps.

**7.** Lift the front half of the snowshoe hare bundle, and draw it rearward. Take 3 to 5 tight thread wraps around the hook shank, tightly abutting the base of the bundle.

**8.** Finish the head of the fly around the hook shank beneath the hair bundle, directly behind the hook eye. Clip the front bundle of snowshoe hare, following the angle of the hook eye, to make a tuftlike head. Clip the wing hairs so that they extend just to the rear of the body.

**9.** Clip a V in the hackle beneath the shank, removing the barbs from an arc of about 90 degrees, as shown in this front view.

# Pale Morning Dun Spinner Patterns

## PMD HARE SPINNER

| Hook: | #14-18 standard dry-fly |
|---|---|
| Thread: | Tan 8/0 |
| Tail: | Light dun Microfibetts |
| Body: | Rusty brown or pale yellow dubbing |
| Wing: | Light dun snowshoe hare's foot hair |

Except for the hook size range and component colors, this fly is identical to the BWO Snowshoe Dun and can be tied using the instructions on pp. 13–15. This highly durable pattern is easy to spot on the surface. Because it floats extremely well, it makes a particularly good choice for fishing broken or choppy water. The fly pictured here is tied with a rusty brown body, but the pattern can be dressed with a yellow body instead, which also makes an excellent dun imitation. Trout can be selective to color during a PMD spinnerfall, so carrying both shades of this pattern is good insurance.

## PMD SNOWSHOE SPENTWING

| Hook: | #14-18 standard dry-fly |
|---|---|
| Thread: | Tan 8/0 |
| Tail: | Light dun Microfibetts |
| Body: | Fine pale yellow or rusty brown dubbing |
| Wing: | Light dun snowshoe hare's foot hair |

Ordinarily, we like a more visible, higher-profile fly, such as a parachute style or the snowshoe Compara-dun type, for fishing spinnerfalls. But there are occasions—extremely flat water or fussy, hard-fished trout—when a flush-riding spent-wing design becomes necessary. A spinner pattern tied with spent, poly yarn wings is commonly used, but we prefer this version tied with snowshoe hare's foot hair. We find that it floats better, and because the hare's foot hair is a bit unruly, it makes a slightly fluffier wing that's easier to see on the water. This is a fairly easy fly to tie, and it's worth carrying both the yellow-bodied pattern pictured here (top view) and a rusty brown version.

**1.** Mount the tying thread behind the hook eye. Lay a thread foundation to the midpoint of the shank, and then return the thread halfway to the hook eye. Prepare and mount a bundle of hare's foot hair as described for the BWO Snowshoe Dun, steps 1–6, pp. 13–14. Take additional tight crisscross wraps over the hair to secure it firmly, crosswise atop the shank, as shown in this top view. Trim off any wayward hairs that distort the wing profile.

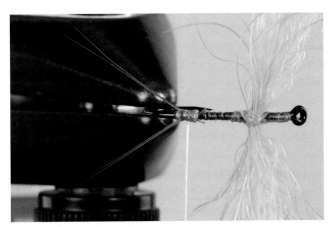

**2.** Using 2 or 3 Microfibetts per side, form a split tail as long as the hook shank, as described for the *Baetis* CDC Compara-dun, steps 3–7, pp. 11–12, and shown in this top view.

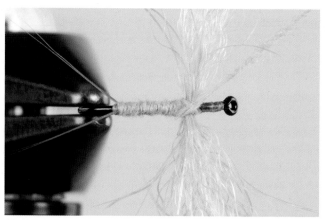

**3.** Using a thinly dubbed thread, form a slender, slightly tapered body up to the hook eye. When you reach the wing, take a couple of crisscross wraps of dubbing between the wings to conceal the wing-mounting wraps. Dub the head of the fly, and tie off the thread behind the hook eye.

**5.** If desired, you can trim away any stray hairs on the bottom of the wings so that the body lies flush in the film. Here, the left wing appears just as it was tied. On the right wing, the stray hairs have been trimmed flush with the body. Leaving the stray hairs on the upper side of the wings intact makes the pattern more visible.

**4.** Raise the wings vertically, and trim to about the length of the hook shank.

# Small Western Green Drake

Despite their name, these aren't exactly small mayflies—the larger imitations are tied on #12 hooks—but they are still smaller than their bigger brethren, the Western Green Drakes, which are often tied on #10 hooks. The Small Western Green Drakes actually comprise two species, *flavilinea* and *coloradensis*, which share a name because the nymphs are difficult to tell apart, live in the same type of water, and exhibit similar behaviors. And though the duns of the two species may differ a bit in size and color, in general they look sufficiently alike that anglers call both of them by the same name. More important, the resemblance is strong enough that the same fly patterns can be used for all the fishable stages of both species. The *flavilinea*, however, tend to emerge earlier in the season and often later in the day, whereas the *coloradensis* typically emerge later in the season and earlier in the day. The nymphs of both species are found in cold streams of all sizes that have moderate to fast flows and fairly clean rock and gravelly bottoms—the type of water ordinarily found at higher elevations. Emergence can continue for weeks, with daily hatches often lasting one or two hours. Even though hatches can be sparse on some waters, it doesn't diminish their importance; these meaty mayflies are a mouthful. Moreover, the best hatches generally occur in the fall, when few other insects are emerging, and trout seize the opportunity to fatten up for the winter on any of these big mayflies they run across.

| | |
|---|---|
| **Other common names:** | Flav (*D. flavilinea*); Slate-Winged Olive, Fall Green Drake (*D. coloradensis*) |
| **Family:** | Ephemerellidae |
| **Genus:** | *Drunella* |
| **Species:** | *flavilinea* and *coloradensis* |
| **Emergence:** | *flavilinea*: Late June to September, late afternoon to evening; *coloradensis*: September, midday |
| **Spinnerfall:** | Midmorning to evening |
| **Body length:** | *flavilinea*: 1/4–3/8 inch (7–9 mm); *coloradensis*: 3/8–7/16 inch (9–11 mm) |
| **Hook sizes:** | #12-16, with #12-14 most common |

**Above:** *Small Western Green Drake nymph: dark brown, olive-brown, or reddish brown.*

**Above:** Flavilinea *dun: dark olive-brown to light olive body with yellow tones; light gray to dark gray wings.*

**Top left:** Coloradensis *dun: olive-brown with yellow-tan or light olive banding on abdomen; light gray to dark gray wings.*

**Left:** *Small Western Green Drake spinner: olive-brown to reddish brown body; clear wings.*

# Important Fishing Stages

**Nymph** Immature nymphs spend most of their time crawling among the rocks in moderate to fast currents. Because their stout, strong legs give a firm grip, the nymphs are well adapted to withstanding turbulent water. They are not often found drifting freely in the current, and blind-fishing a nymph imitation is not particularly productive. When the nymphs mature, as indicated by their dark wing pads, they migrate into slower water and wait until they are ready to hatch. As emergence commences, the nymphs often make a false start or two, swimming upward a short distance and then drifting back to the bottom, before finally making their way to the surface. At this time, just before and during the hatch, nymph imitations can be quite effective. The speed and depth of the water generally dictate how to present a nymph pattern. In waters deeper than 3 feet, when no rising trout are visible, fish the nymph, with added weight, below an indicator. In shallower water, or wherever you spot rises or boils, fish the nymph as a dropper 10 to 20 inches behind a dry fly.

**Emerger/dun** The duns emerge from their nymphal shucks just below or on the surface, and the transformation is generally not a rapid one. The protracted emerger phase of the insect makes emerger patterns a good bet for these mayflies. Likewise, the duns typically spend a fair amount of time on the surface before becoming airborne, and even then they tend to make a few false starts before finally flying off. In the longer runs and pools of larger streams, the length of time it takes the duns to emerge and fly off gives the trout ample opportunity to intercept the insect and the angler a good chance to intercept the trout. But on the short runs and pools of small streams, the emerging duns are often swept downstream into turbulent waters, where many of them are crippled and drown. In waters of this type, fishing a swamped dun or sunken emerger pattern can be productive.

**Spinner** Mating swarms generally occur around riffles. After mating, the males fall spent to the ground or water. The females fly to streamside foliage for a short time, then return to the riffles to deposit their eggs by dipping to the surface a few times, and finally fall spent on the water. Trout feed on these spinners, though generally not in the riffles but in the slower currents of pools, tailouts, and eddies below the faster water. As with virtually all mayfly spinners, it's very difficult to see these spent flies on the water unless you are very close to them, and the problem is compounded when the hatch is sparse and there aren't many flies to see. It can be easy to overlook one of these spinnerfalls, so keep alert for it whenever there has been a hatch. Sometimes the best clue simply comes from trout rising to something you can't see on the water.

# Small Western Green Drake Nymph Pattern

**GOLD-RIBBED HARE'S EAR**

| | |
|---|---|
| **Hook:** | #12-16 standard nymph |
| **Weight (optional):** | Lead or nontoxic wire |
| **Thread:** | Brown 6/0 |
| **Tail:** | Hare's mask guard hairs |
| **Rib:** | Small gold tinsel |
| **Abdomen:** | Medium-brown hare's ear dubbing |
| **Wing case:** | Turkey tail-feather section |
| **Thorax:** | Dark brown hare's ear dubbing |

Universally regarded as one of the preeminent all-purpose flies, the Gold-Ribbed Hare's Ear is, in fact, a highly credible imitation of the Small Western Green Drake nymph. The squat, blocky body and the rough, spiky dubbing suggest the general profile and pronounced legs of the natural insect. These crawler-type nymphs are weak swimmers, so the pattern is best fished on a dead drift. Use a weighted fly below an indicator before or during the hatch; in shallower water or when rises are visible, fish the unweighted version as a dropper behind a dry fly. The colors of the natural nymph vary; using a mixture of hare's ear dubbings—natural brown and dyed olive, for instance—is an option for tiers seeking a closer color match for local populations of the insect.

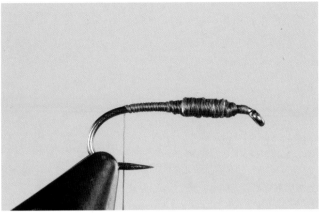

**1.** If tying a weighted fly, as shown here, weight the hook as described for the Pheasant Tail Nymph, steps 1–2, p. 21. Position the thread at the tailing point. For an unweighted fly, mount the thread behind the hook eye, and wrap a thread foundation rearward to the tailing point.

**2.** Clip a bundle of guard hairs from a hare's mask, and mount it atop the shank to make a tail about one hook gap in length.

**3.** Clip the excess tail material. Mount a 3-inch length of tinsel atop the hook shank at the tailing point. Apply dubbing sparsely to the thread, and dub a slightly tapered abdomen to the midpoint of the shank.

**4.** Wrap the tinsel forward with 3 to 5 wraps in an open spiral. Secure the tinsel at the front of the abdomen, and clip the excess. Cut a section of turkey feather one hook gap in width.

**5.** Position the feather atop the shank, with the thinner end projecting rearward and the front, or most distinctly marked, side of the feather facing downward. Mount the feather atop the shank directly in front of the dubbed abdomen. Dub a thorax slightly larger in diameter than the abdomen, stopping 4 to 5 thread wraps' distance behind the hook eye. Position the thread directly in front of the thorax.

**6.** Fold the feather over the top of the thorax, and secure it atop the shank. Use a dubbing needle to pick out hairs on each side of the thorax; trim away any excessively long fibers. Then clip the excess feather, and finish the head. Coat the wing case with head cement for added durability.

# Small Western Green Drake Emerger Patterns

## CDC LOOP WING EMERGER

| Hook: | #12-16 standard dry-fly |
|---|---|
| Thread: | Olive 8/0 |
| Tail: | Brown Antron yarn fibers |
| Rib: | Brown 3/0 thread |
| Abdomen: | Olive or olive-brown Antron dubbing |
| Wing: | 1 or 2 dark dun CDC feathers |
| Thorax: | Olive or olive-brown Antron dubbing |
| Legs: | Barbs from wing feathers |

Though this pattern floats flush in the surface film, it gets fairly good buoyancy from both the CDC and the loop-wing construction. This construction not only puts more CDC into the pattern than would a flatter wing case, but also traps air in the bubble-like wing. When tying larger hook sizes, choose a CDC feather with relatively long barbs in order to produce a smooth, uniform loop wing. If using rather sparse CDC feathers, stack two of them so that the curvatures match and the tips are aligned; treat them as a single feather when you tie, as shown in the following sequence. To fish the fly in the film, treat it with a powder-type floatant; to fish it awash in or just beneath the surface, leave the fly untreated. Because this pattern rides low, it can be difficult to see, especially in choppy water; you can improve strike detection by fishing it as a dropper behind a dry fly.

**1.** Mount the thread behind the hook eye, and wrap a thread foundation rearward to the tailing point. Mount a bundle of Antron fibers to form a trailing shuck about one shank length long. Clip the excess. Mount a 6-inch length of the ribbing thread atop the rearmost tailing wrap. Position the thread at the rear of the fly.

**2.** Dub a slightly tapered abdomen to the midpoint of the shank. Wrap the ribbing thread forward 3 to 5 turns in an open spiral. Secure the ribbing thread at the front of the abdomen.

**3.** Clip the excess ribbing material. Strip away any short or scraggly barbs from the base of the CDC feathers. Stack the feathers so that the tips are aligned and the curvatures match. Position the feathers atop the shank, with the concave side facing upward and the butts pointing toward the rear of the fly. Take 2 firm, but not tight, thread wraps around the feathers, near the butt ends, to mount them directly ahead of the abdomen.

**4.** With your right fingers, put slight tension on the bobbin. With your left fingers, carefully pull the feather butts rearward, sliding the feathers beneath the thread wraps until the tips projecting beyond the thread wraps are about one shank length long.

**5.** Secure the feathers with additional tight wraps toward the hook eye. Clip away the feather tips, and position the thread at the front of the abdomen. Dub a thorax, stopping 5 to 6 thread wraps' distance behind the hook eye. With your left hand, hold a dubbing needle crosswise over the middle of the thorax. With your right fingers, fold the CDC feathers over the dubbing needle to make a loop about one hook gap in height.

**6.** Remove the dubbing needle. Hold the loop wing in place with your left fingers, and secure the feathers directly in front of the thorax.

**7.** Clip the excess feather. Working at the rear base of the wing, clip a few fibers from the outside edges of the wing to form legs on either side of the body, as shown in this bottom view. Finish the head of the fly.

## COMPARA-EMERGER

*Originators: Al Caucci and Bob Nastasi*

| | |
|---|---|
| **Hook:** | #12-16 standard dry-fly |
| **Thread:** | Olive 8/0 |
| **Tail:** | Dun Antron yarn fibers or hen-hackle barbs |
| **Body:** | Olive-brown or olive dubbing |
| **Wing:** | Gray deer hair |
| **Head:** | Butts of wing hair |

This pattern, an emerger version of the well-known Compara-dun, is simplicity itself. It requires few materials, ties up quickly, and is easy to dress. The deer-hair wing makes the fly easy to spot on the surface, gives good flotation even in choppy water, and provides a reasonable color match for the wings on Small Western Green Drakes. In calmer waters, treat only the wing with floatant so that the body rides in or just beneath the surface film, like that of the emerging natural. In broken water, treat the whole fly for better flotation. If there's any trick to tying this pattern, it lies in using the proper amount of deer hair for the wing: too little and the fly won't float; too much and it becomes difficult to mount the hair securely. It takes a bit of experimentation, since deer hair varies somewhat in thickness and compressibility.

**1.** Mount the thread behind the eye, and wrap a thread foundation to the tailing point. Mount the Antron yarn or a bundle of aligned hackle barbs atop the hook to make a tail about one shank length long. With a sparsely dubbed thread, form a slightly tapered body over the rear $^4/_5$ of the hook shank. Position the thread at the front of the body.

**2.** Clean and stack a bundle of deer hair as shown in the instructions for the Sparkle Dun, steps 1–4, pp. 27–28. As a starting point, use the suggestion given in step 1 for sizing the deer-hair bundle. After the hair is stacked, withdraw it from the stacker, and position it as shown so that the hair extends beyond your fingertips about $^3/_4$ the length of the body.

**3.** Hold the hair in position, and pinch it with your left fingers. Secure the hair to the top of the shank with 5 to 7 tight thread wraps. As you wrap, use your left fingers to hold the hair firmly atop the shank and resist the thread torque that would otherwise draw the hair down the far side of the shank.

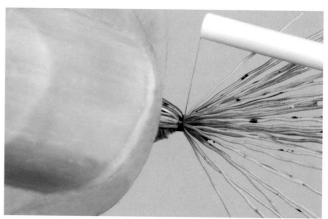

**4.** To further secure the wing, draw the rearmost $^1/_3$ of the hair butts rearward. Take a very tight thread wrap directly in front of these drawn-back hairs.

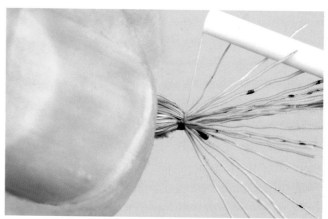

**5.** Draw the next $^1/_3$ of the hair butts rearward, and place a tight thread wrap ahead of them, just as in the previous step.

**6.** Then draw all the hair butts up vertically, and take 4 to 6 tight thread wraps around the shank directly against the base of the butts. Finish the head of the fly around the shank behind the eye. Then clip the hair butts parallel to the hook-eye angle to form a tuftlike head.

# Small Western Green Drake Dun Patterns

## LITTLE GREEN DRAKE

| Hook: | #12-16 standard dry-fly |
|---|---|
| Thread: | Olive 8/0 |
| Tail: | Dun Microfibetts |
| Rib: | Yellow 3/0 thread |
| Body: | Olive-brown dubbing |
| Wings: | Dun snowshoe hare's foot hair |
| Hackle: | Dun dry-fly |

Our preferred selection of flies for Western super hatches contains relatively few standard, collar-hackled patterns. Generally speaking, we find that parachute patterns or hackleless flies give better results. But we make an exception for this Little Green Drake pattern, which works well—possibly because it presents a big, tall-riding profile much like the silhouette of the natural. This fly floats well even when fishing rough water or plunge pools in small streams and can easily support a subsurface dropper. It's also quite durable. Trimming the hackle barbs beneath the shank, as shown in the following sequence, makes a more stable fly that reliably lands and floats upright.

**1.** Mount the thread behind the hook eye. Wrap a thread foundation to the midpoint of the shank; return the thread halfway to the hook eye. Prepare a bundle of snowshoe hare's foot hair, and mount a wing as described for the BWO Snowshoe Dun, steps 1–7, pp. 13–14. When you arrive at step 7 of

that sequence, take 4 tight thread wraps around the base of the wings to secure them upright. If desired, you can produce a more distinct division between the wings by taking crisscross wraps between the near and far wings, as shown in this front view. But the fly will fish just fine without this added step. When the wings are secure, draw them upward with your left fingers, and trim them to about one shank length in height.

**2.** Wrap the thread to the rear of the shank. Mount a bundle of 8 to 10 Microfibetts to form a tail about one shank length long. Clip the excess. Mount a 4-inch length of the ribbing thread, and position the tying thread at the rearmost tail-mounting wrap.

**3.** Dub a slightly tapered abdomen over the rear half of the hook shank. Wrap the ribbing forward 3 to 5 turns in an open spiral over the abdomen. Tie off the ribbing, and clip the excess. Position the tying thread directly in front of the abdomen.

**4.** Strip about one hook shank's length of barbs from the base of a hackle feather. Position the feather edgewise as shown, with the glossy side facing down. The bare hackle stem crosses the hook shank at a 45-degree angle on the near side, directly in front of the abdomen.

**5.** Flatten the thread by spinning the bobbin counterclockwise (as seen from above). Wrap the thread forward, binding the bare hackle stem to the near side of the shank. As you approach the wings, take a firm wrap around the hackle stem tight against the rear base of the wings. Make the next wrap tight against the front base of the wings.

**6.** Take 2 to 3 more tight thread wraps forward. Clip the excess hackle stem, and position the thread 3 to 4 thread wraps' distance behind the hook eye.

**7.** Grasp the hackle in your fingers or a pair of hackle pliers. Take one wrap of hackle directly against, but not overlapping, the abdomen. Continue wrapping the feather forward, keeping the shiny side facing the rear of the hook; each wrap of the feather should abut but not overlap the previous one. When you reach the wings, take a wrap of hackle as close as possible to the rear base of the wings. Complete this wrap with the feather held straight downward beneath the shank.

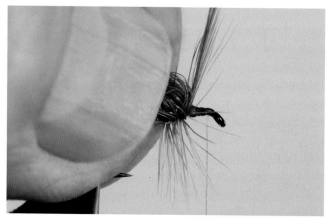

**8.** Use your left fingers to preen the wings rearward, and take a wrap of hackle against the front base of the wings.

**9.** Continue wrapping to the hanging thread. When you reach the thread, hold the hackle feather vertically above the shank. If necessary, unwrap the thread until it is right next to the last wrap of hackle. Transfer the hackle pliers to your right hand, and angle the feather tip forward. With your left hand, take a tight wrap of tying thread over the hackle stem directly ahead of the last hackle wrap.

**10.** Release the hackle tip, and take 2 to 3 tight wraps forward, binding down the feather. Clip the excess feather, and finish the head. Trim the barbs on the underside of the fly so that they are even with the hook point.

## HAIRWING DUN

*Originator: René Harrop*

| | |
|---|---|
| **Hook:** | #12-16 standard dry-fly |
| **Thread:** | Olive 8/0 |
| **Tail:** | Dun Microfibetts |
| **Body:** | Olive dubbing |
| **Rib:** | Yellow 3/0 thread |
| **Hackle:** | Dun dry-fly |
| **Wing:** | Dun deer hair |

This René Harrop pattern is popular for matching a number of mayfly hatches, but we like it best for the Small Western Green Drakes, since the deer hair suggests the darkish wings of the natural insect. Because the hackle beneath the fly is trimmed, the body sits close to the surface film, making a productive pattern in smooth, calm water. But the wing gives the fly good flotation in broken or choppy currents, enough buoyancy to support a nymph dropper, and good visibility to the angler. Though the down-wing style on this fly is normally associated with caddis patterns, it nicely reproduces the bold, rearward-slanting wings of this sizable insect. Except for the wing material, this fly is tied in exactly the same way as the PMD Hare-Wing Dun on pp. 31–32, so the following instructions are abbreviated.

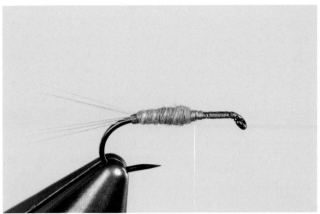

**1.** Mount the thread behind the hook eye, and wrap a thread foundation to the tailing point. Using 2 to 4 Microfibetts per side, form split tails as long as the hook shank, as described for the *Baetis* CDC Compara-dun, steps 3–7, pp. 11–12. Mount a 4-inch length of ribbing thread, and position the tying thread at the tailing point. Dub a slightly tapered abdomen to the midpoint of the shank. Wrap the ribbing thread forward 3 to 5 turns in an open spiral over the abdomen; tie off the ribbing ahead of the dubbing, and clip the excess. Position the thread directly ahead of the abdomen.

**2.** Mount the hackle feather, dub the thorax, and wrap and tie off the hackle as described for the PMD Hare-Wing Dun, steps 2–5, pp. 31–32.

**3.** Clean and stack a bundle of deer hair as shown in the instructions for the Sparkle Dun, steps 1–4, pp. 27–28. Position the hair above the shank so that the tips just reach the rear of the abdomen.

**4.** Mount the wing and finish the fly as shown in the instructions for the Compara-emerger, steps 3–6, p. 40.

**5.** Clip a V in the hackle beneath the shank, removing the barbs from an arc of about 90 degrees, as shown in this front view.

# Small Western Green Drake Spinner Pattern

**BROWN PARACHUTE SPINNER**

| | |
|---|---|
| **Hook:** | #12-16 standard dry-fly |
| **Thread:** | Brown 8/0 |
| **Wing post:** | Light gray or yellow Antron yarn |
| **Hackle:** | Light dun dry-fly |
| **Tail:** | Light dun Microfibetts |
| **Body:** | Fine brown dubbing |

As noted earlier, parachute patterns make excellent low-floating spinner imitations and have the advantage of an elevated wing post for visibility. This fly is generally easy to spot on the surface, but if you have difficulty seeing it, you can dress the pattern with a yellow wing post, which really stands out on the water. In our experience, however, bright wing posts can spook trout—not always, and possibly not even most of the time, but sometimes. Still, if a bright wing post allows you to track the drift of the fly and detect a strike, the trade-off may be worth it. As a compromise, you can trim a bright wing post while you are fishing, cutting it down so that you can just make it out on the water and reducing its conspicuousness to the trout. And as with other parachute patterns, this one can be dressed with a dark colored wing post for better visibility in flat light or overcast conditions. This pattern is tied using exactly the same steps shown for the Parachute BWO on pp. 15–17.

After the fly is completed, if you wish, you can trim away the hackle barbs projecting over the hook eye in an arc of about 90 degrees, as shown in this top view. This gap simplifies threading the fly on the leader and suggests the separation of spent wings.

# Trico

Trico nymphs are found in streams or parts of streams with slow flows and silty, weedy bottoms. Depending on the water conditions and geographic location, Tricos can have one or more generations annually, but on most Western streams they usually produce two generations a year, with the second brood emerging as the first hatch cycle ends. Tricos from the first generation are generally larger than those that follow. The daily and seasonal duration of the emergence, the overlapping stages of the hatch, and the vast number of flies on the water make this truly a super hatch. These same factors can also make trout highly choosy when feeding on Tricos, so it's important to understand the progression of the hatch and the feeding sequence of the fish. The hatch begins with the trout feeding on the female nymphs, emergers, and duns; then the male spinners; and finally the female spinners. During the middle of a long hatch, you may find female emergers, female duns, and both male and female spinners all on the water at the same time. Trout may selectively feed on one of these stages while ignoring the others. Since it is virtually impossible to visually differentiate among these tiny insects from a distance, you may have to change flies until you find the one that matches what the trout prefer.

| Other common names: | Trike, White-Winged Curse |
| --- | --- |
| **Family:** | Leptohyphidae |
| **Genus:** | *Tricorythodes* |
| **Emergence:** | Mid-June to October, with August to mid-September the most common; males during the night, females from sunrise to midmorning |
| **Spinnerfall:** | Early morning to midday |
| **Body length:** | ⅛–¼ inch (3–6 mm) |
| **Hook sizes:** | #18-22 |

**Above:** *Trico nymph: light brown to dark brown or olive-brown abdomen; dark brown thorax.*

**Left:** *Trico nymph (underside): lighter shades of abdomen color.*

**Top left:** Female Trico dun: green abdomen and dark brown thorax; light dun wings.

**Left:** Female Trico dun (underside): light shades of top color.

**Above:** Male Trico dun: dark brown with light dun wings.

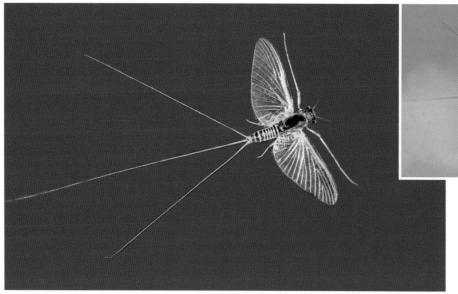

**Left:** *Male Trico spinner: dark brown to black body; clear wings. Insects that fall spent with abdomens extended, such as this one, show light colored banding.*

**Above:** *Male Trico spinner (underside): same as top color.*

**Left:** *Female Trico spinner: pale green to dark green abdomen and dark brown to black thorax; clear wings.*

**Above:** *Female Trico spinner (underside): lighter shade of top color.*

# Important Fishing Stages

**Nymph** Except during an emergence, Trico nymphs are seldom found drifting in the current. When they mature, they either crawl up weeds or drift to the surface to hatch. Male Trico nymphs emerge at night, so they are not important to anglers. But in the early morning, before the females begin to emerge and for a short time after they are hatching, a Trico nymph pattern fished close to the bottom can be productive. During the hatch, most anglers prefer to fish dry-fly patterns, but a nymph imitation fished close to the surface behind a small indicator or dry fly is often very effective on heavily fished streams.

**Emerger/dun** The male duns emerge at night and generally don't produce a fishable hatch. The female duns, however, typically emerge from early to midmorning. During heavy hatches, trout often feed selectively on either the female duns or the emergers and cripples, which may be floating or submerged. Observing a riseform can give you a clue about where to present a fly. A small bubble left in the middle of the rings indicates a dun, emerger, or cripple taken from the surface; no bubble suggests subsurface feeding to a nymph, emerger, or cripple. An emerger pattern, floating or sunken, fished as a dropper behind a more visible dun pattern is a good opening gambit early in the hatch for targeting trout that are not rising to duns. The duns normally spend a fair amount of time on the water and are relatively easy to spot, but often their numbers are so great that it's difficult to find your fly among all the naturals. You can use a brightly colored wing or wing post to help find your fly, but wary trout sometimes reject patterns with bright wings. Probably the best way to keep track of your fly and catch trout during this hatch is to place the fly about 6 inches directly in front of the feeding trout. It requires accurate casting, and you may spook a fish or two, but by delivering the fly to a small target area, you'll be able to pinpoint it on the surface and see the strike.

**Spinners** In the early morning hours, male Trico duns molt into spinners and form a mating swarm that often looks like a cloud of smoke over or adjacent to the water. The females molt shortly afterward, join the swarm, and couple with the males in flight. After mating, the spent males fall to the water. The females fly back to streamside foliage for a short time to extrude their eggs onto the tips of their abdomens. Then they fly back to the water, dip to the surface, release their eggs, and fall spent to the surface. These spinner-falls often occur en masse, and the large number of spent spinners—beginning with the males and ending with the females—will bring up large trout to feed. Spent, flush-floating Trico spinners are difficult to see on the water unless they drift quite close to you. During a hatch, if no insects are visible where rising trout are leaving small bubbles behind, chances are the fish are taking spinners. Carefully search the water to find a natural, and match your pattern to its size and color. Avoid casting to a group of rising trout in the hope that one of them will find your fly; instead, pick a specific fish and place the fly directly in front of it.

# Trico Nymph Pattern

| Hook: | #18-20 1XS nymph |
| Weight (optional): | Lead or nontoxic wire |
| Thread: | Dark brown 8/0 |
| Tail: | 3 pheasant tail-feather barbs |
| Rib: | Fine copper wire |
| Abdomen: | Olive-brown or dark brown dubbing |
| Wing case: | Dark brown Antron yarn fibers |
| Thorax: | Olive-brown or dark brown dubbing |
| Legs: | Antron yarn from wing case |

As indicated in the list of materials, a short-shank hook is best suited for imitations of this small, compact nymph. The dark brown Antron suggests the darkened wing case on a mature nymph. Female Trico nymphs tend to have a lighter body color than the males, so before and during the hatch, fish the olive-brown version of this fly close to the bottom under an indicator. The fly is quite light, so weight should be added to the leader. Adding weight to the fly is also an option, but it won't make the fly sink very fast; there isn't room on the shank to add much wire. But added weight will make the fly penetrate the surface film more easily for fishing as a dropper behind a dry fly once the hatch begins.

**1.** Mount the thread behind the hook eye, and wrap a foundation to the midpoint of the shank. Mount a 4-inch length of copper wire at the midpoint of the shank. Wrap the thread rearward to the tail-mounting point, binding the wire to the near side of the hook shank. Align the tips of 3 pheasant tail barbs, and mount them atop the shank to form tails one hook gap in length. Clip the excess, return the thread to the tail-mounting point, and dub a slightly tapered abdomen to the midpoint of the shank.

**2.** Counterwrap the copper wire forward (counterclockwise when viewed from the hook eye). Take 3 to 4 turns of the wire in an open spiral; tie off the wire in front of the abdomen, and clip the excess. Mount a strand of Antron yarn atop the shank directly in front of the abdomen. Clip the excess, and position the thread at the front of the abdomen.

**3.** Dub a thorax slightly larger in diameter than the abdomen, stopping about 3 thread wraps' distance behind the hook eye. Fold the yarn smoothly over the top of the thorax, and secure it directly in front of the thorax.

**4.** Fold 3 to 5 yarn fibers back along each side of the body. Secure them with 2 to 3 thread wraps taken over the fold to slant the fibers rearward, as shown in this top view. Clip these fibers to about one hook gap in length. Then trim the excess yarn over the hook eye, and finish the head.

# Trico Emerger Patterns

## CDC FEMALE TRICO EMERGER

| Hook: | #18-22 standard dry-fly |
|---|---|
| Thread: | Dark brown 8/0 |
| Tail: | Cream Antron yarn fibers |
| Abdomen: | Fine olive dubbing |
| Wing: | Dun CDC barbs |
| Thorax: | Fine dark brown dubbing |

This emerger design is simple to tie even in the tiny hook sizes sometimes required for late-season Tricos. Since the CDC emerging wing is trimmed relatively short, this fly is not an outstanding floater, and a powder-type floatant is recommended for fishing the pattern dry. But the fly fishes quite well when awash in or just under the surface film, in which case strikes are most easily detected if it's used as a dropper behind a dry fly. We carry several of these flies on the water, since they are easily contaminated by fish slime and algae and will eventually refuse to float. At that point, rinse the fly in clear water, blow away the excess, and leave it on a fly patch or in a fly box to dry. Avoid the temptation to overdress this pattern; the naturals are small and trim.

**1.** Mount the thread behind the hook eye, and wrap to the tailing point. Mount a sparse bundle of Antron fibers atop the shank to form a trailing shuck about one hook shank long. With a sparsely dubbed thread, dub a slightly tapered abdomen over the rear ²/₃ of the shank. Position the thread at the front of the abdomen.

**2.** Prepare a bundle of CDC barbs as described for the BWO Puff Fly, steps 1–3, p. 9. Mount the midpoint of the bundle directly ahead of the abdomen so that the tips of the CDC barbs point rearward.

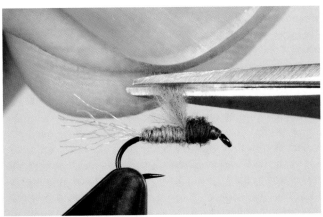

**3.** Trim the butts of the CDC barbs, and secure them with additional thread wraps. Return the thread to the rearmost wing-mounting wraps. With a sparsely dubbed thread, form the thorax to the hook eye. Finish the head of the fly. Lift the bundle of CDC fibers vertically, and trim it to about ²/₃ of a shank length in height.

## KLINKHAMER FEMALE TRICO

| | |
|---|---|
| **Hook:** | #18-22 fine-wire scud |
| **Thread:** | Brown 8/0 |
| **Tail (optional):** | Tan Antron yarn fibers |
| **Abdomen:** | Fine olive dubbing |
| **Wing post:** | Light dun poly yarn |
| **Thorax:** | Fine dark brown dubbing |
| **Hackle:** | Dun dry-fly, one size larger than normally used for the hook size |

The Klinkhamer design is becoming justifiably popular in this country, and this version is a good imitation of an emerging or stillborn female dun in the surface film. Unlike many other Trico emerger patterns, this one is relatively easy to see on the water because of the light colored wing post. A darker-colored post, such as dark dun, can be used when fishing in the silvery glare of low light. Antron fibers imitating the nymphal shuck can be added to give the pattern a more realistic look, as shown below, but the fly works well without them. Dress only the wing post and hackle with a paste-type floatant; the body of the fly will ride just beneath the surface film like that of an emerging natural.

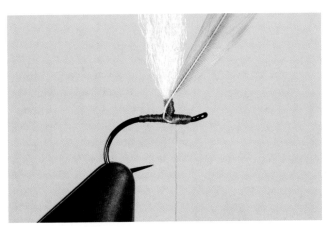

**1.** Mount the tying thread behind the hook eye, and wrap a thread foundation over the front half of the shank. Position the thread in the middle of this foundation. Mount the wing post and hackle using the procedure shown for the Parachute BWO, steps 1–4, pp. 15–16.

**2.** Wrap the thread to the tailing point. Mount a sparse bundle of Antron fibers to form a trailing shuck the length of the hook shank. Dub the abdomen over the rear $^1/_2$ of the shank. Position the thread at the front of the abdomen.

**3.** Dub a thorax slightly larger in diameter than the front of the abdomen, stopping 3 to 4 thread wraps' distance behind the hook eye. Position the thread at the front of the thorax.

**4.** Wrap the hackle around the wing post, and tie it off in front of the thorax as described for the Parachute BWO, steps 7–10, pp. 16–17.

**5.** Trim the hackle tip and finish the head. Draw the wing post upward, and trim it to about one shank length in height.

# Trico Dun Patterns

**SNOWSHOE FEMALE TRICO COMPARA-DUN**

| | |
|---|---|
| **Hook:** | #18-22 standard dry-fly |
| **Thread:** | Brown 8/0 |
| **Tail:** | Dun Microfibetts |
| **Abdomen:** | Fine olive dubbing |
| **Wing:** | Medium dun snowshoe hare's foot hair |
| **Thorax:** | Fine dark brown dubbing |

This winging style does produce some bulk in the thorax and abdomen areas, but it's appropriate for this small, thick-bodied insect. And the winging technique allows you to take advantage of material on the hare's foot that is too short for some of the other snowshoe hare winging techniques presented in this book. Avoid overdressing the wing on this fly; snowshoe hare is a superb dry-fly material, and it doesn't take much of it to float the tiny hooks used for this imitation. This pattern is surprisingly easy to spot on the water for its size, has exceptional buoyancy, and is quite durable. If the wing becomes matted with algae or fish slime, swish the fly in clear water, blot it dry, and treat it with any type of floatant. This same pattern tied with a light dun wing is a good spinner imitation; the male spinner can be tied with a dark brown or black body.

**1.** Mount the thread behind the hook eye, and wrap a thread foundation to the midpoint of the shank. Position the thread ⅓ of a shank length behind the eye. Clip and clean a bundle of snowshoe hare's foot hair as described for the BWO Snowshoe Dun, steps 1–2, p. 13. Mount the bundle, hair tips facing forward, tightly atop the shank.

**2.** Trim the butts of the wing hair at an angle, slanting toward the hook bend, and secure them with thread. Draw the tips of the hair rearward, and take several tight thread wraps abutting the base of the hair to post the bundle upright.

**3.** Wrap the thread to the tailing point, and mount one Microfibett atop the shank to form a tail 1½ shank lengths long. Position the thread at the rearmost tail-mounting wrap.

**4.** Form a thread bump, and mount one Microfibett on each side of the shank as described for the *Baetis* CDC Comparadun, steps 3–7, pp. 11–12. All 3 tails should be of equal length, as shown in this top view. Trim the excess tail material.

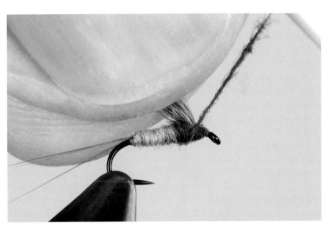

**5.** Dub a tapered abdomen to the midpoint of the shank. Dub the thorax to the rear of the wing. Take a crisscross wrap of dubbing beneath the wing to conceal the wing-mounting wraps.

**6.** Dub to the hook eye, and finish the fly. Raise the wing vertically, and trim it to about one shank length in height.

**7.** Preen the wing fibers into a 180-degree arc, as shown in this front view. Put a drop of head cement at the base of the wing to lock the fibers in position.

## PARACHUTE FEMALE TRICO DUN

| Hook: | #18-22 standard dry-fly |
|---|---|
| Thread: | Brown 8/0 |
| Tail: | Dun Microfibetts |
| Abdomen: | Fine olive dubbing |
| Wing post: | Medium dun or light dun poly yarn |
| Hackle: | Dun dry-fly |
| Thorax: | Fine brown dubbing |

Some version of a parachute pattern is standard operating equipment for the Trico hatch, for the simple reason that the design works so well. That the poly yarn used for the wing post comes in a variety of colors is an added advantage. The medium dun indicated in the recipe is a good match for the wings of the natural, but if you have trouble seeing it on the water, try a light dun or white post or, in low-light conditions, a dark brown or even black post. We usually carry this pattern tied with wing posts of two or three different colors to maximize the visibility of this small fly under a variety of light conditions. Keep the pattern relatively sparse. Three or four wraps of high-quality hackle is all it takes to float a small hook. This pattern can also be used to imitate the female Trico spinner or, if dressed with a dark brown or black body, the male spinner. Since the dressing steps are virtually identical to those for the Parachute BWO, pp. 15–17, we have abbreviated the instructions here.

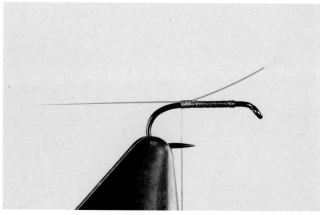

**1.** Mount the thread about one hook eye's width behind the eye, and wrap a thread foundation to the rear of the shank. Mount one Microfibett atop the shank to form a tail 1½ shank lengths long. Position the thread at the rearmost tail-mounting wrap.

**2.** Form a thread bump, and mount one Microfibett on each side of the shank as described for the *Baetis* CDC Compara-dun, steps 3–7, pp. 11–12. All 3 tails should be of equal length, as shown in this top view.

**3.** Trim the excess tail material. Dub a tapered abdomen to the midpoint of the shank. Mount a yarn post and hackle feather as described for the Parachute BWO, steps 1–4, pp. 15–16.

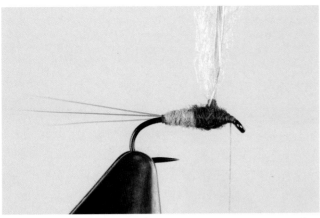

**4.** Dub a thorax slightly thicker than the abdomen, stopping 3 to 4 thread wraps' distance behind the hook eye. Position the thread at the front of the thorax.

**5.** Wrap and tie off the hackle as explained in steps 7–10, pp. 16–17. Trim away the feather tip, and finish the head of the fly. Trim the wing to about one shank length in height.

# Trico Spinner Patterns

## PARACHUTE TRICO SPINNER

| | |
|---|---|
| **Hook:** | #18-22 standard dry-fly |
| **Thread:** | Black 8/0 |
| **Tail:** | Light dun Microfibetts |
| **Male body:** | Fine black dubbing |
| **Female body:** | Abdomen: fine light olive dubbing; thorax: fine dark brown dubbing |
| **Wing post:** | White poly yarn |
| **Hackle:** | Light dun dry-fly |

This pattern is our top choice for a Trico spinnerfall. Though the wing post is clipped short to be less conspicuous to the trout, it is still visible enough to spot on the water. If you have difficulty seeing the fly, try an orange poly yarn post, clipped short to leave just a stub above the uppermost hackle wrap. Apply a drop of head cement just above the top hackle wrap, and let it bleed down the post to secure the hackle. As with all small parachute flies, avoid the tendency to over-hackle; three or four wraps is generally sufficient. The fly can be treated with either a paste or powder floatant and can be tied to imitate either a male or female spinner. The male spinner, shown in the photo, also makes a good adult midge imitation. The steps for dressing this fly are identical to those shown for the Parachute Female Trico Dun, pp. 52–53, except that the wing post is trimmed to about ½ a shank length in height.

After the fly is completed, if you wish, you can trim away the hackle barbs projecting over the hook eye in an arc of about 90 degrees. This gap simplifies threading the fly on the leader and suggests the separation of spent wings.

## POLY-WING SPINNER

| Hook: | #18-22 standard dry-fly |
|---|---|
| Thread: | Black 8/0 |
| Tail: | Light dun Microfibetts |
| Male body: | Fine black dubbing |
| Female body: | Abdomen: fine light olive dubbing; thorax: fine dark brown dubbing |
| Wing: | White poly yarn |

Poly yarn is a highly practical wing material, particularly on smaller flies. The material is simple to work with, and controlling the volume of the finished wing is somewhat easier with poly yarn than with the snowshoe hare's foot hair shown elsewhere in this book. This pattern essentially uses a Compara-dun-type wing formed of poly yarn, and the arc-wing design, though not strictly imitative of the spent wings on a spinner, does not usually seem to bother the trout. The wing style is significantly easier to see and floats better in choppy water than wings that lie flat against the surface. The male spinner imitation is pictured here; the fly can also be tied to imitate the female.

**1.** Mount the thread about one hook eye's width behind the eye, and wrap a thread foundation to the rear of the shank. Form the triple tail as described for the Parachute Female Trico Dun, steps 1–2, p. 52, and shown in this top view. Position the thread at the rearmost tail-mounting wrap. Dub a slightly tapered abdomen to the midpoint of the shank.

**2.** Clip a 2-inch length of poly yarn, and separate out a bundle of fibers suitable for the desired density of the finished wing. The yarn is mounted just like the bundle of snowshoe hare's foot hair shown in the instructions for the BWO Snowshoe Dun, steps 4–7, p. 14.

**3.** Dub the thorax to within 2 to 3 thread wraps' distance of the hook eye, and finish the head of the fly. Draw both wings upward, and trim to one shank length in height.

**4.** Use your fingers or a dubbing needle to spread and separate the wing fibers and distribute them evenly into a narrow fan shape, as shown here from the front. Apply a drop of head cement to the base of the wing fibers to lock them in position.

# Western Green Drake

The Western Green Drakes comprise three species that are important to anglers: the *grandis* and *doddsi*, which hatch in early summer, and the *spinifera*, which hatches in late summer. Of all the major mayfly super hatches, the Western Green Drake is arguably the most fickle. Pinning down the emergence dates can be difficult to begin with, and even then the already short hatch duration—typically about a week—may have only one or two peak days. But even sparse emergences of these big, beefy mayflies bring up large trout. Streams that have fishable hatches of this mayfly are well known, but the capricious emergence can vary from year to year and day to day, depending on weather and water conditions. When you hit it right, however, this hatch offers some first-rate emerger and dry-fly fishing; even when only a few duns are on the water or when the hatch has been over for a week or more, you can cast a Green Drake pattern to likely holding water and still take trout. During the summer months, especially on higher-elevation streams, it's always wise to have at least one or two Green Drake emerger and dun patterns in your vest, just in case you run across this hatch; few other flies that trout anglers ordinarily carry will match the size and color of these insects.

| | |
|---|---|
| **Other common name:** | Green Drake |
| **Family:** | Ephemerellidae |
| **Genus:** | *Drunella* |
| **Emergence:** | June to August, late morning to midafternoon |
| **Spinnerfall:** | After dark |
| **Body length:** | ½ to ¾ inch (12–16 mm) |
| **Hook sizes:** | #8-12, with #10 most common |

**Above:** *Western Green Drake nymph (*grandis*): brown with red or olive tones.*

**Top right:** *Western Green Drake nymph (*grandis*, underside): usually a lighter shade of the top color.*

**Right:** *Western Green Drake nymph (*grandis*): this side view shows the pronounced spines on the abdomen.*

**Left:** *Western Green Drake nymph (doddsi): generally light brown to dark brown with olive, yellow, or red tones.*

**Above:** *Western Green Drake nymph (doddsi, underside): usually a lighter shade of the top color; the fringe of gills forms a suckerlike disk on the abdomen.*

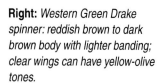

**Left:** *Western Green Drake dun: bright green to dark green body with yellow, olive-yellow, or brown banding; medium gray to dark gray wings, sometimes with yellow-olive tones.*

**Above:** *Western Green Drake dun (underside): same as top color.*

**Right:** *Western Green Drake spinner: reddish brown to dark brown body with lighter banding; clear wings can have yellow-olive tones.*

# Important Fishing Stages

**Nymph** All three species of Western Green Drake nymphs are found on the rocky bottoms of cold trout streams with medium to fast currents. When the nymphs mature, they migrate out of the faster waters into more moderate flows. These stout, sturdy insects are able to crawl through brisk water with ease, so not many of them are normally found in the drift. At emergence, they leave the bottom and swim feebly upward, typically drifting a good distance before reaching the surface. Fishing a nymph pattern close to the bottom before a hatch in waters below riffles and along the seams between fast and moderate flows can be productive, but fishing a nymph as a dropper close to the surface during the hatch is often more effective. The *grandis* and *spinifera* nymphs are quite similar in appearance; the *doddsi*, with its squat, compact body, has a slightly different shape. But the nymphs of the three species share similar habits and habitats, and the variations in body size and shape are minor enough that we haven't found separate imitations to be necessary.

**Emerger/dun** The duns emerge just below or on the surface. It normally takes them a while to slip out of their nymphal shucks, extend and stiffen their wings, flap them tentatively, and lift off from the surface. Most get airborne, but many are trapped in choppy water and eventually drown. The extended time it takes the duns to emerge and the high number of cripples that occur during this hatch make these mayflies easy prey for the fish. Trout are generally not selective during this hatch, but they can become so at times, especially on more popular waters. When fishing over fussy trout, observe the rises to determine whether they are feeding subsurface on nymphs or emergers or taking emergers or duns from the surface. Then select the appropriate type of pattern. The duns of the three species resemble one another in size, shape, and body colors, and the same patterns can be used to imitate them all.

**Spinner** Spinnerfalls occur after dark and are not considered a fishable stage of the life cycle. If a spinnerfall happens very late at night, however, you may see a few trout feeding on spent spinners in eddies in the early morning. Anglers who spend a lot of time fishing Green Drake waters during the hatch season may find that it pays to carry a spinner pattern, but the parachute dun pattern in this section is a workable substitute.

# Western Green Drake Nymph Pattern

**GREEN DRAKE RUBBER LEGS**

| | |
|---|---|
| **Hook:** | #10-12 standard nymph |
| **Weight:** | Lead or nontoxic wire |
| **Thread:** | Dark brown 6/0 |
| **Tail:** | Hareline Brown Buggy Nymph Legs |
| **Rib:** | Copper wire |
| **Abdomen:** | Olive-brown or dark brown dubbing |
| **Wing case:** | Dark brown Antron yarn |
| **Thorax:** | Olive-brown or dark brown dubbing |
| **Legs:** | Hareline Brown Buggy Nymph Legs |

Using thin rubber or latex legs gives this pattern good mobility in the water. It's best to weight this fly only lightly; too much weight on the hook shank can deaden the movement of the fly underwater. Nymph imitations are usually weighted with wire that is approximately the same diameter as the hook shank, but for this pattern, use wire that is $1/2$ to $2/3$ the hook-shank diameter. Before the hatch, fish this fly under an indicator close to the bottom by adding weight to the leader as needed. Present it dead-drift, since these nymphs are weak and clumsy swimmers. When the hatch begins, fish this pattern as a dropper behind a dry fly—an approach that works even when the trout are visibly taking duns.

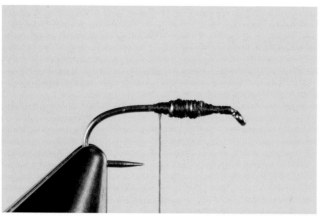

**1.** Mount the thread and weight the hook as described for the Pheasant Tail Nymph, steps 1–2, p. 21. Position the thread at the base of the thread ramp behind the wire underbody.

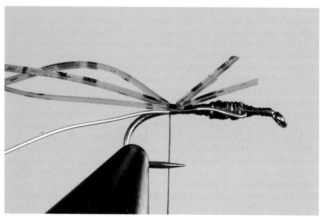

**2.** Mount a 4-inch length of copper wire, and wrap the thread rearward, about halfway to the tail-mounting point, binding the wire to the near side of the hook shank. Mount three 2-inch lengths of the tail material as a bundle atop the shank.

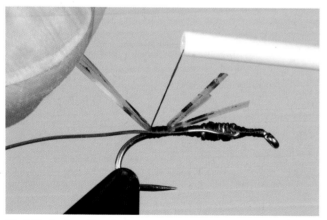

**3.** Lift the bundle of tail material slightly, and stretch it rearward. Wrap the thread rearward to the tailing point on the hook, binding the tails to the top of the shank and the copper wire to the near side of the shank.

**4.** Trim the excess tail and rib materials, and secure with additional thread wraps. Trim the tails to one hook gap in length, as shown in this top view. Dub a tapered abdomen over the rear half of the shank, and position the thread directly ahead of the abdomen.

**5.** Spiral the ribbing wire forward in 4 to 5 wraps over the dubbing. Tie off the wire in front of the abdomen, and clip the excess. Clip a length of Antron yarn, and mount it atop the shank at the front of the abdomen. Position the thread midway between the front of the abdomen and the hook eye.

**6.** Clip two 2-inch lengths of rubber leg material. Hold the tying thread away from you, elevated above the shank at 45 degrees. Fold one strand of leg material around the thread from underneath, as shown in this front view.

**7.** Wrap the thread around the shank until it points directly at you. As you make this wrap, let the thread tension draw the rubber leg against the far side of the shank, as shown in this top view.

**8.** Secure the leg with another thread wrap or two. Mount the near-side leg by pulling the tying thread directly toward you at 45 degrees below the hook shank. Fold the other strand of leg material over the top of the thread, as shown in this front view.

**9.** As in step 7, wrap the thread around the shank, letting the thread tension draw the rubber strand against the near side of the shank. Secure this leg with additional thread wraps, as shown in this top view.

**10.** Dub the thorax. First pull the legs forward to dub between the abdomen and the rear set of legs. Next, dub between the front and rear sets of legs. Finally, pull the legs rearward to dub the front of the thorax, stopping 4 to 6 thread wraps' distance behind the hook eye.

**11.** Fold the wing case smoothly and evenly over the thorax. Secure it directly in front of the thorax. Clip the excess, and finish the head. Trim the legs to be about ½ the length of the hook shank.

# Western Green Drake Emerger Patterns

## CDC OUTRIGGER EMERGER

| | |
|---|---|
| **Hook:** | #10-12 standard dry-fly |
| **Thread:** | Yellow 6/0 or 8/0 |
| **Tail:** | Pheasant tail-feather barbs |
| **Abdomen:** | Dark olive goose or turkey biot |
| **Thorax:** | Dark olive Antron dubbing |
| **Wing:** | 2 CDC feathers |
| **Legs:** | Butt ends of wing feathers |

In the larger hook sizes used for this imitation, this pattern is best tied with CDC feathers that have long, uniform barbs and nicely formed, intact tips. Not all CDC feathers meet these requirements, and we usually sort through a bag of material and set these feathers aside for use on larger CDC flies. If you have only sparser, shorter-barbed feathers, you can use two feathers, stacked so the curvatures match, for each side of the wing. Secure them tightly, then clip away the two extra feather stems before forming the legs. Apply a powder floatant to the fly to keep it in or on the surface film, and fish it is as you would a dry fly. To fish the pattern submerged, omit the floatant, presoak the fly, and fish it as a dropper behind a more visible dun imitation. This is an excellent pattern for calmer waters or selective trout. Except for the body, the dressing instructions are the same as those for the CDC and Biot PMD, pp. 29–31, so the steps are abbreviated here.

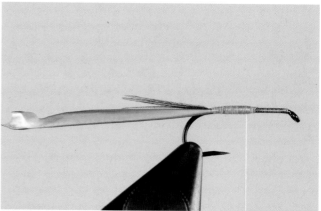

**1.** Mount the tying thread behind the hook eye, and wrap a smooth thread foundation to the tailing point. Align the tips of 3 pheasant tail barbs, and mount them atop the shank to form a tail as long as the hook shank. Do not clip the excess. Position the thread at the rearmost tailing wrap. Strip (do not clip) a biot from a biot stem. Look for the notch at the base of the biot, and mount the biot on the near side of the hook shank with this notch facing upward. Flatten the thread by spinning the bobbin counterclockwise when viewed from above. Bind down the tip of the biot and the pheasant tail butts, forming a smooth, slightly tapered underbody to the midpoint of the shank. Clip the excess tail material. Position the thread at the midpoint of the shank.

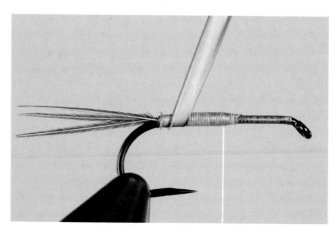

**2.** Clip the tip of the biot in a pair of hackle pliers, and take one wrap around the shank, covering the tail-mounting wraps. Notice the hairlike fringe at the rear of this wrap. Take a second wrap of the biot, overlapping the first, but leaving the fringe visible, as shown here.

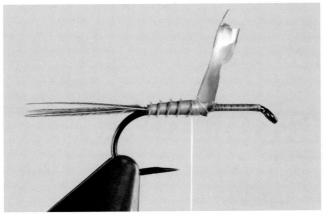

**3.** Continue wrapping the biot forward, with consistent spacing between the wraps, always leaving the fringe of the previous wrap visible. Wrap to the tying thread. Secure the biot, and then wrap the thread rearward with 3 tight wraps.

**4.** Clip the excess biot. Dub a thorax, slightly larger in diameter than the abdomen, halfway to the hook eye. Position the thread at the front of the thorax.

**5.** The wings and legs are dressed as described for the CDC and Biot PMD, steps 4–7, p. 30. Finish the head. Trim the legs to be about ½ the length of the hook shank, as shown in this bottom view.

## GREEN DRAKE QUIGLEY CRIPPLE

*Originator: Bob Quigley*

| | |
|---|---|
| **Hook:** | #10-12 standard dry-fly |
| **Thread:** | Olive 6/0 |
| **Tail:** | Olive-dyed grizzly marabou |
| **Rib:** | 1 strand of yellow floss |
| **Abdomen:** | Olive Antron dubbing |
| **Thorax:** | Peacock herl |
| **Wing:** | Deer hair |
| **Hackle:** | Olive-dyed grizzly dry-fly |

This emerger-cripple design is a good match for many mayfly hatches, but we favor it particularly for the Green Drakes. The larger hooks used for this imitation permit a longer deer-hair tuft that incorporates more of the buoyant portion of the hair shaft into the pattern than smaller hooks do. Though the dressing should be kept sparse, the fly floats well and is quite visible even on choppy or broken water. You can dress the entire fly with powder floatant (paste will mat the marabou tail) for a flush-floating body, or you can squeeze some water into the abdomen and treat only the wing and hackle for a fly that rides with the body angled beneath the surface film. If using three- or four-strand floss, separate out a single strand of the material for ribbing the fly.

**1.** Mount the thread behind the hook eye, and lay a thread foundation to the tailing point. Align the tips of 10 to 12 marabou barbs, clip them from the stem, and mount them atop the shank to make a tail about one shank length long. Clip the excess. Mount a 4-inch strand of the ribbing floss atop the rearmost tailing wraps, and clip the excess. Position the thread at the rearmost tail-mounting wrap.

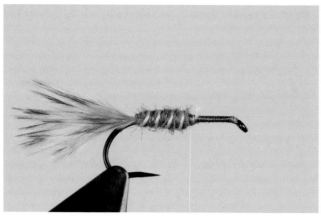

**2.** Dub a tapered abdomen to the midpoint of the shank. Spiral the ribbing floss forward in 4 to 6 wraps over the abdomen. Tie off and clip the excess. Position the thread directly in front of the abdomen.

**3.** Align the tips of 4 or 5 peacock herls along with a 6-inch length of scrap tying thread, and trim away the uppermost ½ inch of the bundle. Secure the herl and tying thread atop the shank directly ahead of the abdomen. Position the thread midway between the abdomen and hook eye. Draw the herls and thread downward as a bundle. Clip the end of the bundle in a pair of hackle pliers, and gently twist the bundle 3 or 4 times; don't be too aggressive, or the fragile herls will break. Wrap the bundle forward to the tying thread, pausing to retwist it as necessary to maintain a chenille-like strand.

**4.** Secure the peacock herls, and clip the excess. Clean and stack a bundle of deer hair as described for the Sparkle Dun, steps 1–4, pp. 27–28. Mount it atop the shank so that the tips projecting forward of the mounting wraps are as long as the hook shank. Between the butts and tips, create a smooth thread foundation that is 4 to 5 thread wraps wide. Position the tying thread at the rear of this foundation.

**5.** Trim the deer-hair butts to be even with the rear of the thorax. Prepare and mount a hackle feather as described for the Little Green Drake, steps 4–6, p. 42. Wrap the thread forward, lift the hair-tip wing, and take one wrap of thread around the shank ahead of the wing.

**6.** Take 4 to 5 wraps of hackle over the thread foundation as described for the Little Green Drake, steps 7–8, p. 42.

**7.** When you reach the wing, draw the deer hair rearward, and take one wrap of hackle around the hook shank directly ahead of the wing. Then secure the hackle tip as described for the Little Green Drake, steps 9–10, pp. 42–43. Clip the excess and finish the head of the fly.

# Western Green Drake Dun Patterns

## FOAM-BODIED GREEN DRAKE PARACHUTE

| | |
|---|---|
| **Hook:** | #12 TMC 206BL |
| **Thread:** | Yellow 6/0 |
| **Tail:** | 3 dun-colored paintbrush bristles |
| **Body:** | Green Larva Lace Dry Fly Foam |
| **Wing post:** | Deer hair |
| **Hackle:** | Olive or medium dun dry-fly |
| **Cement:** | Dave's Flexament, or similar flexible cement, thinned 1-to-1 with solvent |

This flush-floating fly works extremely well in calm or choppy waters; it is easy to spot on the water and can be used as an indicator for a trailing emerger pattern. Despite how it looks, this is not a difficult fly to tie, in the technical sense, and it's quite durable. It does require more steps than other patterns, but in our experience, the effectiveness of this fly makes the extra effort worthwhile. Instead of the Larva Lace, you can substitute a strip of a similar 1/16-inch (2 mm) closed-cell foam. For information on types of foam suitable for this pattern, see p. 79. Tying the foam body extension is identical to tying the body on the Buoyant Brown Drake dun pattern, p. 79, so these steps are abbreviated in the following sequence.

**1.** Though this pattern can be tied with tails of moose mane or even Microfibetts, we prefer to use paintbrush bristles. They are larger in diameter and stiffer than Microfibetts, which makes them easier to work with, and they are significantly more durable than moose mane. Look for suitable fibers on a small trim brush; the bristles should taper smoothly to a point rather than having blunt or rounded ends.

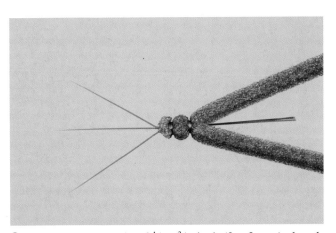

**2.** Using a 3-inch strip of 1/8 x 3/32-inch (3 x 2 mm) closed-cell foam, form the extended body and tails as described for the Buoyant Brown Drake, steps 1–6, pp. 80–81. For this fly, however, only 2 body segments are required. After forming the second segment, half-hitch the thread wraps forming the segment, and complete the tail as shown in steps 7–8, p. 81.

**3.** Place a hook in the vise. Mount the thread behind the hook eye, and wrap a foundation to the rear of the shank. Then wrap forward to the point where the thread hangs 3 thread wraps' distance in front of the hook point. The wraps laid to the rear of the hanging thread are necessary to provide a secure foundation for mounting the body extension.

**4.** Clean and stack a bundle of deer hair as described for the Sparkle Dun, steps 1–4, pp. 27–28. This hair has no appreciable function in floating the fly, so use only enough to create a distinct wing silhouette. Too much hair creates excessive bulk on the shank. Mount the hair atop the hook shank so that the tips extend one hook length in front of the mounting wraps. Trim the butt ends to a taper, and bind them down. Position the thread at the frontmost mounting wrap.

**5.** Draw the deer hair rearward, and take several tight thread wraps abutting the front of the hair bundle to raise the wing vertically. Take 12 to 14 firm, but not extremely tight, thread wraps up the bundle of hair to form a thread foundation. Then take an equal number of wraps down the bundle, and position the thread in front of the wing post.

**6.** Prepare and mount the hackle feather as described for the Parachute BWO, steps 3–4, p. 16.

**7.** Wrap the thread to the rearmost thread wrap on the hook shank. Apply cement to a short section of the rearmost thread wraps. Fold the foam extension around the sides of the hook shank so that the length of foam behind the hanging thread is a little longer than the adjacent segment on the body extension. The butts of the tail material should lie along the hook shank, sandwiched between the foam halves.

**8.** Secure the foam to the hook shank with 4 tight thread wraps.

**9.** Fold the foam strips rearward. Clip the tail butts, and wrap the thread to the rear base of the wing post.

**10.** Fold the foam strips forward and pull them slightly downward so that the tops are just above the hook shank. Then use 4 thread wraps to secure them.

**11.** Fold the strips back, and wrap the thread forward to a point about 4 thread wraps' distance behind the hook eye. Fold the strips forward and downward, as in the previous step, and secure the foam with 4 tight thread wraps.

**12.** Trim the excess foam, forming a neat head. Tie off the thread, but do not clip it. Instead, draw the thread rearward over the top of the frontmost body segment, and take 2 thread wraps around the base of the wing post and hackle. Position the thread directly behind the wing post, atop the wraps previously taken to form the body segment.

**13.** Take 5 to 6 wraps of hackle down the wing post as described for the Parachute BWO, steps 7–8, p. 16. Finish by holding the hackle feather on the near side of the shank, angled slightly downward toward the rear of the fly. Take 4 tight thread wraps horizontally around the base of the wing post and over the feather tip to secure the hackle. When the hackle tip is secure, bring the thread forward over the top of the frontmost body segment. Take 3 thread wraps behind the hook eye, and finish the head of the fly. Trim away the feather tip, and apply cement to the thread wraps.

## GREEN DRAKE WULFF

| Hook: | #10-12 standard dry-fly |
| --- | --- |
| Thread: | Green 6/0 |
| Tail: | Moose body hair |
| Rib: | 1 strand of yellow floss |
| Body: | Green dubbing |
| Wings: | Deer hair |
| Hackle: | Dark dun dry-fly |

Like all Wulff-style flies, this one is admirably suited to fast, choppy water; it rides high and handsomely and stands out quite visibly on the surface. Perhaps surprisingly, though, this pattern is also a good producer on smooth, flat water. Because it floats well, it's an ideal fly from which to trail a nymph or emerger pattern. During a sparse hatch, and on the days following a hatch, use this pattern to search likely holding waters. Its bold and conspicuous presence seems to trigger strikes. As with all hair-wing flies, finding just the right amount of hair for the wings takes a bit of experimentation. It's best to err on the sparse side; too much hair is difficult to secure to the shank and creates excessive bulk. We think that trimming the hackle on the underside of the shank gives better stability to the fly, but this step can be omitted if desired. If using three- or four-strand floss, separate out a single strand of the material for ribbing the fly.

**1.** Mount the thread behind the hook eye. Wrap a tight foundation to the midpoint of the shank, and position the thread ⅓ of a shank length behind the hook eye. Clean and stack a bundle of deer hair as described for the Sparkle Dun, steps 1–4, pp. 27–28. Mount the bundle as explained in steps 5–7, p. 28.

**2.** Draw the deer hair rearward, and use tight thread wraps to build a tapered ramp or cone of thread abutting the front of the hair bundle to raise the wing vertically. Keeping this thread ramp smooth will help in wrapping the hackle later.

**3.** Use a dubbing needle and your fingers to divide the wing fibers into 2 equal bundles. Take 3 or 4 very firm crisscross wraps between the two wings to separate them, as shown in this top view.

**4.** Take 3 or 4 wraps of thread under light tension around the base of each wing to consolidate the fibers, as shown in this top view. Adjust the wings as necessary to make them symmetrical about the hook shank and separated by an angle of 45 to 60 degrees.

**5.** Wrap the thread to the tail-mounting point. Clip a bundle of moose body hair that is about ¼ of a hook gap in thickness. Clean and stack it as described for the Sparkle Dun, steps 1–4, pp. 27–28. Mount the hair atop the shank to form a tail one shank length long. Clip the butts at an angle, as shown. When the excess tail material is bound down, it will mate with the wing butts to form a smooth underbody.

**6.** Bind down the excess tail material. Return the thread to the tailing point, and mount a 4-inch length of ribbing floss. Dub the abdomen to the midpoint of the shank. Take 4 to 5 wraps of floss in an open spiral over the dubbing; tie off the rib directly ahead of the abdomen, and clip the excess.

**7.** Prepare and mount a hackle feather as shown in the instructions for the Little Green Drake, steps 4–6, p. 42. Position the thread 3 to 4 thread wraps' distance behind the eye.

**8.** Wrap the hackle and tie it off as described for the Little Green Drake, steps 7–10, p. 42–43. Clip the excess feather, and finish the head.

**9.** If desired, trim the hackle barbs beneath the shank to be even with the hook point, as shown in this front view.

# Western March Brown

Because March Brown duns normally emerge before spring runoff, they have the distinction of being the first sizable mayfly of the season to make an appearance—frequently in numbers sufficient for the trout to take notice, and sometimes in very dense hatches. The hatch can go on as long as six weeks, with the daily emergence lasting one to three hours, depending on the weather. During cool, wet conditions, the hatches tend to be drawn out and intermittent; you may see a modest number of duns on the water and sporadic rises. In milder weather, the emergence is condensed into a shorter time period that concentrates the duns on the surface and tends to bring up more feeding trout. During the hatch, the nymphs, emergers, and duns drift a fair distance downcurrent, either below or on the surface, and the length of time they spend in and on the water makes them easy prey for trout. Though the early-season weather can make for miserable fishing conditions, the March Brown hatch is a favorite among anglers, because the flies are large and offer some of the first good dry-fly fishing of the year.

| | |
|---|---|
| **Other common name:** | March Brown |
| **Family:** | Heptageniidae |
| **Genus:** | *Rhithrogena* |
| **Emergence:** | March to May, late morning to late afternoon |
| **Spinnerfall:** | Late afternoon to evening |
| **Body length:** | ¼–⅝ inch (6–15 mm) |
| **Hook sizes:** | #12-14 |

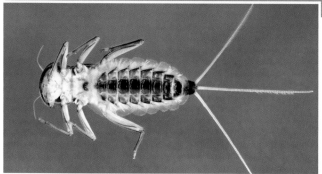

**Top:** *Western March Brown nymph: brown, often with shades of olive or red.*

**Bottom:** *Western March Brown nymph (underside): same color as top, with a lighter thorax; gills around the perimeter of the abdomen form a suction disk.*

**Top:** *Western March Brown dun: shades of brown, olive, or gray; mottled brown wings.*

**Bottom:** *Western March Brown dun (underside): lighter shade of top color.*

**Right:** *Western March Brown spinner: brown to reddish brown on top, tan to light tan on underside; clear wings.*

# Important Fishing Stages

**Nymph** The nymph of this mayfly lives in the faster flows of riffles and runs. With its flattened body and large, overlapping gills that can form a suction cup on the bottom of the abdomen, the March Brown nymph can easily crawl around on the bottom of brisk currents. The only time these nymphs are readily available to trout is during the hatch—first when they migrate out of the faster flows into adjacent moderate currents where the emergence takes place, and then when they are making their way to the surface to hatch. During their trip to the surface, the nymphs tend to drift downstream a good distance. Fish a nymph pattern close to the bottom in riffles and nearer to the surface in the waters below them.

**Emerger/dun** The duns emerge on or just below the surface. The length of time it takes the dun to emerge and fly off the water varies with the weather: the cooler it is, the more time it takes. Floating emerger patterns work extremely well for this hatch, especially in calmer runs and below riffles. In choppier water, use a dun pattern that offers a little more flotation and visibility. Under most light conditions, these large duns are easy to spot as they ride on the water or fly above it. If you see numerous duns on the surface and in the air, but few or no rises, it's a fairly reliable sign that the trout are taking nymphs close to the bottom. This behavior is not uncommon during a March Brown hatch; just keep fishing a nymph pattern until the trout show on the surface, or move along the stream until you find rising fish.

**Spinner** The female spinners lay their eggs in the faster flows of riffles and runs, and then fall spent on the water; most sink quickly beneath the surface. March Brown spinners are generally not considered important enough to imitate. But if no duns are in the air or on the water, and you notice boils from fish feeding close to the surface, chances are that the trout are taking sunken spinners. A soft-hackle fly is a good imitation of a sunken spinner; fish it drag-free, by itself or behind an indicator or dry fly.

# Western March Brown Nymph Pattern

**PHEASANT TAIL NYMPH**

*Originator: Frank Sawyer*

| | |
|---|---|
| **Hook:** | #12-14 standard nymph |
| **Weight (optional):** | Lead or nontoxic wire |
| **Thread:** | Brown 6/0 or 8/0 |
| **Tail:** | Pheasant tail-feather barbs |
| **Rib:** | Fine copper wire |
| **Abdomen:** | Pheasant tail-feather barbs |
| **Wing case:** | Pheasant tail-feather barbs |
| **Thorax:** | Peacock herl |
| **Legs:** | Pheasant tail-feather barbs |

Though a number of patterns have been specifically designed to match the nymph of the March Brown mayfly, day in and day out we think a standard Pheasant Tail Nymph works as well as any of them. It has the added virtues of credibly imitating other mayfly species and being a good searching pattern. This fly is tied using the instructions for the Pheasant Tail Nymph on pp. 21–23. You can vary the color of the abdomen by using a mix of natural brown and dyed-olive pheasant tail barbs.

The Western March Brown nymph has a wide body for its length, and you can suggest that breadth by building a tapered underbody of tying thread beneath the abdomen after the tails and ribbing are mounted, as shown here. If the pheasant tail barbs prove too short to wrap the entire abdomen, wrap one bundle of barbs as far as it will go, then trim the excess, mount a second bundle, and finish the abdomen.

Here's the finished fly; note the short, stocky profile.

# Western March Brown Emerger Patterns

## MARCH BROWN WARREN EMERGER

*Originator: Gary Warren*

| | |
|---|---|
| **Hook:** | #12-14 standard dry-fly |
| **Thread:** | Brown 6/0 |
| **Tail:** | Mottled brown hen-hackle barbs |
| **Abdomen:** | Olive-brown dubbing |
| **Wing case:** | Deer hair |
| **Thorax:** | Tan dubbing |
| **Wing:** | Deer-hair tips from wing case |

Even during a strong hatch with lots of duns on the water, we probably catch more trout on floating emerger patterns than on adult imitations—overwhelmingly so early in the hatch, when most of the feeding is subsurface, and again as the hatch winds down, when stillborns and cripples are mostly what remain on the water. And this is our go-to pattern in March Brown season. Treat the whole fly with floatant. This pattern fishes well presented to trout on a dead drift, but it is also effective when swung in wet-fly fashion; the forward-slanting arc wing creates a wake in the water, much like a waking steelhead fly. For some reason, this rather unnatural behavior strongly appeals to the trout. To get the most buoyancy in the fly, choose deer hair with a uniform-diameter shaft that tapers abruptly to a short tip, rather than tapering gradually to a long, threadlike tip.

**1.** Mount the thread behind the hook eye, and wrap a thread foundation to the tailing point. Align the tips of 6 to 8 hen-hackle barbs, and mount them atop the shank to form a tail one hook gap in length. Trim the excess, and dub a slightly tapered body to the midpoint of the shank. Position the thread at the front of the abdomen.

**2.** Clean and stack a bundle of deer hair about ½ the hook gap in thickness as described for the Sparkle Dun, steps 1–4, pp. 27–28. Mount the bundle directly in front of the abdomen so that the hair tips extend to the tip of the tail. Clip and bind down the butts. Position the thread at the front of the abdomen.

**3.** Dub a thorax slightly larger in diameter than the abdomen, stopping 5 to 6 thread wraps' distance behind the hook eye.

**4.** Fold the deer hair smoothly and evenly over the top of the thorax. Keep all the hair on top of the shank, and secure it directly ahead of the thorax with tight thread wraps.

**5.** Draw the hair tips rearward, and form a tapered head against the base of the hair to elevate the wing anywhere between 45 and 90 degrees. Finish the head of the fly behind the hook eye, around the hook shank only.

**6.** Fan the hair outward into an arc of 180 degrees over the top of the shank, as shown in this front view.

## MARCH BROWN FLYMPH

*Originator: Rick Hafele*

| | |
|---|---|
| **Hook:** | #12-14 standard nymph |
| **Thread:** | Red Pearsall's Gossamer Silk |
| **Tail:** | Pheasant tail-feather barbs |
| **Rib:** | Fine gold tinsel |
| **Body:** | Dark hare's ear dubbing |
| **Hackle:** | Brown hen |

This version of Pete Hidy's wingless wet-fly style is a little simpler to tie than the original but just as effective. The materials on this pattern give the fly good mobility and translucence in the water. Use this fly when duns are on the water but you see only subsurface boils caused by trout taking nymphs or emerging duns just beneath the film. You can treat this fly with a powdered floatant and fish it dead-drift on the film, or leave it untreated to fish just beneath the surface. Perhaps the most effective technique, however, is the standard wet-fly method: cast quartering downstream, and let the fly swing on a tight line directly in front of a feeding fish. The body of this fly should be roughly dubbed. Don't twist the dubbing into a tight yarn on the thread; twist it only loosely so that the dubbing fibers splay and give a brushlike appearance to the body.

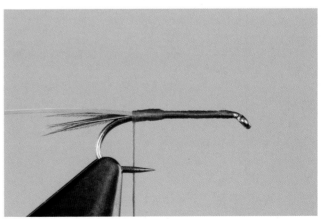

**1.** Mount the thread behind the hook eye, and wrap a foundation to the tailing point. Align the tips of 4 to 6 pheasant tail barbs, and mount them to form a tail one hook gap in length. Cut a 4-inch strip of ribbing tinsel, and mount it atop the rearmost tail-mounting wrap.

**2.** Dub a tapered body over the rear ⁴/₅ of the hook shank. Position the thread at the front of the body. As noted above, the body should be roughly dubbed. If the dubbed body is rather smooth in appearance, use a dubbing needle or teaser to pick out some fibers around the perimeter of the body. Spiral the tinsel forward in 4 to 5 wraps over the body. Tie off the tinsel directly ahead of the body, and clip the excess.

**3.** Position the thread directly ahead of the body. Prepare and mount the hackle as shown in the instructions for the Little Green Drake, steps 4–6, p. 42, except that the hackle here should be mounted with the top, or shiny, side of the feather facing upward. Position the thread 3 to 4 thread wraps' distance behind the hook eye.

**4.** Clip the tip of the feather in a pair of hackle pliers, and take one wrap around the shank directly in front of the body.

**5.** With your left fingers, draw the wrapped hackle barbs rearward, and take a second wrap of the feather.

**6.** Tie off the hackle as described for the Little Green Drake, steps 9–10, pp. 42–43. Clip the feather tip. Draw all the hackle barbs rearward, and form a tapered head that slightly overlaps the base of the last hackle wrap to slant the hackle barbs toward the rear of the fly. Then finish the head of the fly.

# Western March Brown Dun Patterns

## MARCH BROWN COMPARA-DUN

*Originator: Al Caucci and Bob Nastasi*

**Hook:** #12-14 standard dry-fly
**Thread:** Tan 6/0
**Tail:** Tan Microfibetts
**Body:** Tan dubbing
**Wing:** Deer hair

Though this simple, easily tied pattern floats well, it sits flush on the surface and presents a well-defined wing silhouette from every angle, making it particularly effective on light riffles and flat, glassy currents. This Compara-dun style is our top choice for imitating the March Brown adults. The deer hair is a credible match for the mottled wings of the dun. The tan dubbing used here may look a bit light in color compared with the natural insect, but once wet, the dubbing gets a little darker. But more important, the dubbing is selected to match the color of the underside of the dun, which is noticeably lighter than the top of the body. The fly is tied using the procedure shown for the Sparkle Dun, pp. 27–29, except that split tails are used in the place of the trailing shuck, so the instructions here are abbreviated.

**1.** Dress the wing as shown in the instructions for the Sparkle Dun, steps 1–10, pp. 27–29, and shown here from the front.

**2.** Use one Microfibett per side to dress split tails one shank length long, as described for the *Baetis* CDC Compara-dun, steps 3–7, pp. 11–12, and shown here in a top view.

**3.** Apply dubbing sparsely to the thread to form a slim body. When you reach the base of the wing, take a crisscross wrap of dubbing underneath the shank to cover the wing-mounting wraps, as shown in this bottom view. Dub the head, and finish the fly.

## HAIRWING MARCH BROWN

*Originator: René Harrop*

| Hook: | #12-14 standard dry-fly |
|---|---|
| Thread: | Tan 8/0 |
| Tail: | Tan Microfibetts |
| Body: | Tan dubbing |
| Hackle: | Medium dun dry-fly |
| Wing: | Deer hair |

With its rearward-slanting wing and thorax hackle, this pattern gives a more realistic suggestion of the March Brown dun than does the Compara-dun style. In the hook sizes specified in the list of materials, the fly contains plenty of the buoyant shafts of deer hair. So while the pattern rides low in the water by design, it floats well and is a good choice for fishing broken currents or using with a nymph or emerger dropper attached. Except for the wing material, this fly is tied using the same procedure as the PMD Hare-Wing Dun, pp. 31–32, so the following steps are abbreviated.

**1.** Dress the tails, body, and hackle as shown in the instructions for the PMD Hare-Wing Dun, steps 1–5, pp. 31–32.

**2.** Clean and stack a bundle of deer hair about ½ the hook gap in thickness, as described for the Sparkle Dun, steps 1–4, pp. 27–28. Mount the bundle atop the shank with 6 to 7 tight thread wraps, directly in front of the hackle, so that the hair tips extend to the end of the body.

**3.** Finish securing the wing and trimming the excess deer hair as described for the Compara-emerger, steps 3–6, p. 40. Trim the hackle beneath the hook shank to leave gap of about 90 degrees, as shown in this front view.

# Western March Brown Spinner Pattern

**SOFT-HACKLE SPINNER**

*Originator: Sylvester Nemes*

| | |
|---|---|
| **Hook:** | #12-14 standard nymph |
| **Thread:** | Orange or hot orange 8/0 |
| **Tail:** | White Antron yarn |
| **Rib:** | Fine gold wire |
| **Body:** | Rusty brown dubbing |
| **Hackle:** | Cream hen |
| **Head:** | Rusty brown dubbing |

Though soft-hackle flies have been used for centuries, Montana tier Syl Nemes was their foremost advocate and popularizer in recent decades. The design is usually associated with sunken flies and down-and-across wet-fly presentations, but it gives excellent results when treated with a powder-type floatant (paste will mat the hackle barbs) and fished dead-drift on or in the surface film. The soft barbs collapse on the water, creating the impression of spent spinner wings. The sparse dressing and low profile can make this fly dreadfully hard to see on the water, and it's often fished on a dropper behind a more visible dry fly. This is a very simple pattern to tie, but avoid the tendency to overhackle this fly; a couple of wraps will do the job.

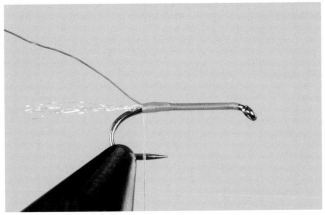

**1.** Mount the thread behind the eye, and wrap a foundation to the tailing point. Mount a sparse bundle of Antron fibers atop the shank to form a tail one shank length long. Mount the rib-

bing wire atop the rearmost tail-mounting wrap, and clip the excess materials. Position the thread at the rearmost tail-mounting wrap.

**2.** Using a sparsely dubbed thread, form a slightly tapered body over the rear ¾ of the hook shank. Spiral the wire in 7 to 8 wraps over the body. Secure the wire in front of the body, and clip the excess. Position the thread directly in front of the body.

**3.** Prepare, mount, wrap, and tie off the hackle as shown in the instructions for the March Brown Flymph, steps 3–6, p. 72. Position the thread directly ahead of the hackle.

**4.** With a sparsely dubbed thread, form a tapered head, and then finish the fly.

# LOCALLY IMPORTANT HATCHES

## Brown Drake

Because of its specialized habitat requirements, the Brown Drake is not a common sight in Western trout streams, where the silty, sandy substrate that the nymphs require is the exception rather than the rule. Such waters do exist here and there throughout the region, however, and most, though not all, are well known to fly shops in those areas. But the size of this fly and the readiness with which trout take it can make it among the most important hatches of the year for anglers who fish streams with good Brown Drake habitat. The emergence usually lasts for a week or two, and the best way to determine if the hatch cycle has started is to check the calmer water along the banks or in eddies for floating nymphal shucks. The shucks are large and easy to spot; scoop one up in the palm of your hand with a little water, and it will fill out to the size and shape of the living insect.

| | |
|---|---|
| **Family:** | Ephemeridae |
| **Genus:** | *Ephemera* |
| **Emergence:** | June to August, twilight into the night |
| **Spinnerfall:** | Twilight into the night |
| **Body length:** | Nymphs: ½–¾ inch (12–20 mm); duns: ⅜–⅝ inch (10–15 mm) |
| **Hook sizes:** | #8-12 |

**Above:** *Brown Drake dun: yellow-brown with dark brown markings on top of abdomen; mottled, pale brown wings.*

**Above:** *Brown Drake spinner: yellow-tan to brown with dark markings on top of abdomen; clear wings with brown markings.*

**Top:** *Brown Drake nymph: pale yellow-brown with dark brown markings on top of abdomen.*

**Bottom:** *Brown Drake nymph (underside): shades of yellow-brown with brown mottling on abdomen.*

**Right:** *Brown Drake nymph (swimming): the nymph swims with an up-and-down undulation of its abdomen.*

# Important Fishing Stages

**Nymph** The nymphs of this mayfly dig burrows into silty, sandy, fine-graveled bottoms of streams or sections of streams with slow flows. They leave their shelters only after dark to feed or, eventually, to hatch. They are agile swimmers, using undulating up-and-down movements of their tails and abdomens to propel themselves in short bursts through the water. At emergence, the nymphs leave the bottom and swim to the surface, where the duns crawl from their nymphal shucks on or just below the film. For a month or two leading up to the hatch, after sunset when the nymphs are most active, you can fish a nymph imitation close to the stream bottom. But the most productive time to use a nymph pattern is during an emergence, as the nymphs are swimming toward the surface. Cast quartering upstream, and retrieve the fly with short strips as it drifts down.

**Emerger/dun** Emergence occurs from dusk into the night, though on cloudy days it may occur earlier. The duns need time to unfurl and dry their wings, so they typically spend a fair amount of time on the water. These large mayflies, and the rises of trout feeding on them, are easy to see during early twilight, but after dark, you have to fish the water or, if your directional hearing is reliable, cast to the noisy rises. When fishing at night, keep most of the slack out of your fly line and leader, or you'll miss a lot of strikes. Use at least a 4X tippet to avoid breaking off a trout when it takes. Fish an emerger or dun imitation drag-free, but give it light twitches during the drift; the movement calls attention to the fly and also reduces slack in the line for better strike detection and more hookups.

**Spinner** The spinners typically arrive on the water at about the same time that the duns are emerging. Normally, trout do not feed selectively on one or the other, but it's prudent to carry imitations of both. Most of the time, a dun pattern will entice spinner-feeding trout, but should it fail, a spinner imitation will often do the job.

# Brown Drake Nymph Pattern

**BEADHEAD STRIP NYMPH**

| | |
|---|---|
| **Hook:** | #8-10 standard nymph |
| **Head:** | Gold bead |
| **Weight:** | Lead or nontoxic wire |
| **Thread:** | Brown 6/0 |
| **Tail:** | Hareline gold-variant rabbit strip |
| **Wing case:** | Dark brown Antron yarn |
| **Thorax:** | Gold Arizona Dubbing |
| **Legs:** | Hareline Brown Buggy Nymph Legs |

This is a version of Gary Borger's Strip Nymph. The rabbit-strip tail is highly mobile underwater, and this pattern is designed to be fished with short, quick, 4–6-inch strips or twitches of the rod tip, with each strip or twitch followed by a pause. This continuous strip-and-pause action causes the rear portion of the fly to flip up and down as the fly drifts, mimicking the swimming movements of the natural. This motion is exaggerated by the weighted head, which imparts a jiglike action to the fly.

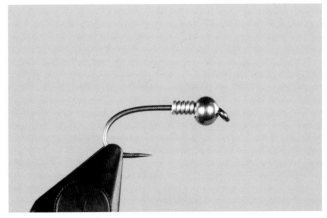

**1.** Position a bead behind the hook eye. Apply 4 to 6 wraps of wire to the hook shank, and then slide the frontmost wraps into the hole at the rear of the bead as shown in the instructions for the Beadhead Prince Nymph, steps 1-2, p. 141.

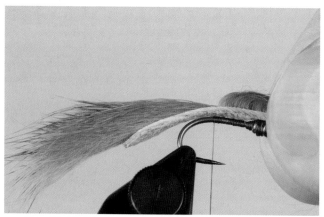

**2.** Mount the thread behind the wire, and secure the wire with thread wraps. Wrap a thread foundation to the rear of the hook shank. At the end of the rabbit strip where the hair tips extend past the hide, trim the hide to a point. Moisten the hair of the rabbit strip, and form a gap in the hair one shank length ahead of the pointed end of the strip. Position this gap over the hanging thread.

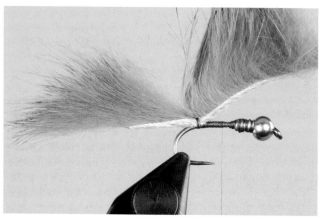

**3.** Use 3 tight thread wraps to secure the hide to the top of the hook shank. Then fold the front of the rabbit strip rearward, and wrap the thread to the midpoint of the hook shank.

**4.** Fold the rabbit strip forward, and form a gap in the hair, as you did in step 2, directly above the hanging thread. Secure the hide to the top of the hook shank with tight thread wraps through the gap in the hair. Trim the excess rabbit strip, and cover the tag end with thread wraps. Position the thread at the rearmost wrap securing the rabbit strip.

**5.** Mount a length of Antron yarn, and trim the excess. Dub the thorax halfway to the rear of the bead.

**6.** Mount the legs on each side of the hook shank as described for the Green Drake Rubber Legs, steps 6–9, pp. 58–59. Secure the front legs with additional forward wraps to position them about 3 thread wraps' distance behind the bead, as shown in this top view.

**7.** Dub the rest of the thorax, stopping directly behind the front legs. Position the thread between the front legs and rear of the bead as shown.

**8.** Fold the yarn over the top of the hook shank, and secure it behind the bead. Trim the excess, tie off the thread behind the bead, and coat the thread wraps with head cement. Trim the legs to about the length of the hook shank.

# Brown Drake Dun Pattern

**BUOYANT BROWN DRAKE**

| Hook: | #12-14 TMC 206BL |
|---|---|
| Thread: | Brown 6/0 |
| Tails: | Tan paintbrush bristles |
| Abdomen: | Tan Larva Lace Dry Fly Foam |
| Cement: | Dave's Flexament, or similar flexible cement, thinned 1-to-1 with solvent |
| Head/wing: | Deer hair |

This is a highly buoyant and durable pattern that, with a few small changes, we favor for imitating emergers, duns, and spinners of larger mayfly species. Despite its size, it's lightweight and easy to cast, and it lands reliably upright. The wing makes this fly relatively easy to see at twilight, but after dark, it's a different story. During a dead drift, give the line an occasional twitch to reduce slack and stay in touch with the fly. Use a similar light twitch whenever you hear the sound of a rise coming from the vicinity of the fly. If a trout has taken the fly, the twitch will help set the hook; if not, you can continue the drift. This kind of movement imparted to a dead-drifting fly during the daytime often causes trout to turn away; at night, it often does the opposite. Despite the number of steps in the fol-

lowing sequence, dressing this fly is not difficult; we've presented the construction of the extended body in detail because it may be unfamiliar to many tiers. We use this foam extended-body style for a number of patterns in this book, so a word about types of closed-cell foam is in order.

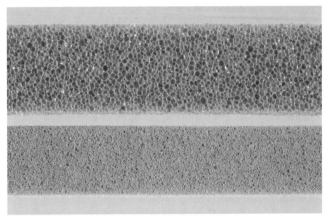

There are different types of closed-cell foam. The type normally sold for fly tying, called craft foam (shown at the bottom), is very dense and finely textured, with tiny chambers that supply flotation. We find this material somewhat heavy, not very flexible, prone to cut under thread pressure, and only moderately buoyant. Flies tied with it ride low in the film. It is, however, quite durable and can certainly be used to tie any foam-bodied fly in this book. It's frequently sold in sheets 2 mm thick, from which you can cut strips 3 mm wide to use on this Brown Drake pattern. It can also be cut to other widths as specified for other patterns in this book.

For most foam-bodied flies, however, we prefer the lighter, coarser-textured, low-density closed-cell foam with larger interior bubbles (shown at the top). This material is much lighter, floats higher, and is somewhat easier to tie. As of this writing, only one brand of this foam is available: Larva Lace Dry Fly Foam. It is used in the following sequence, as well as for nearly all of the foam-bodied flies that appear in this book. Larva Lace foam comes in presliced sheets, with both wide and narrow strips. The narrow strips are the right size for this pattern and for all patterns in this book requiring $\frac{1}{8}$ x $\frac{3}{32}$-inch (3 x 2 mm) strips.

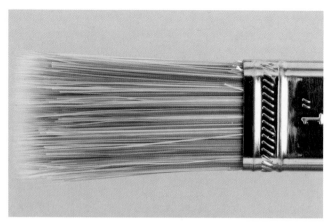

The tails on this pattern are best fashioned from paintbrush bristles. Though similar to Microfibetts and other synthetic tailing materials, paintbrush bristles are thicker and stiffer, making them easier to work with, and they are more durable than moose mane. Select bristles that taper neatly to a point rather than having blunt or broken tips.

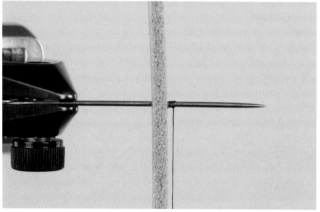

**1.** Mount a sewing needle in the vise. Take a 3-inch strip of ⅛ x ³/₃₂-inch (3 x 2 mm) closed-cell foam, and push the very center of the slightly wider face of the strip onto the needle, as shown in this top view. Mount the thread directly in front of the foam with firm, but not excessively tight, wraps.

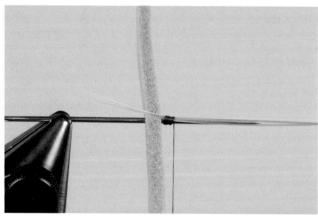

**2.** Align the tips of 3 paintbrush bristles. Using 3 firm, but not tight, thread wraps, mount them atop the needle to form a ½-inch-long tail. Position the thread at the frontmost mounting wrap.

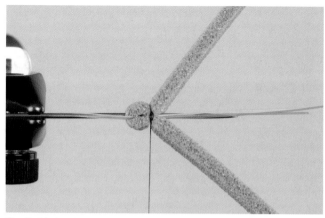

**3.** Fold the 2 ends of the foam forward to sandwich the needle and the bristle butts between them, as shown in this top view. Take 3 snug thread wraps around the foam to form a body segment that is about ⅛ inch in length. Then place 3 half hitches directly over the thread wraps.

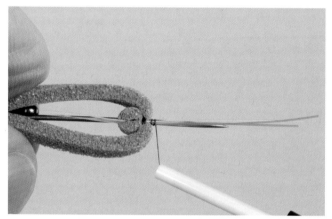

**4.** Fold the ends of the foam rearward, as shown here from the top, and spiral the thread forward about ¹/₁₆ inch.

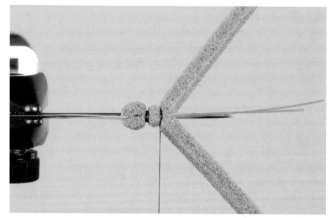

**5.** Fold the foam tags forward again, as shown here from the top, sandwiching the needle and bristles between them. Use 3 thread wraps to create a second segment that is about ¹/₁₆ inch in length.

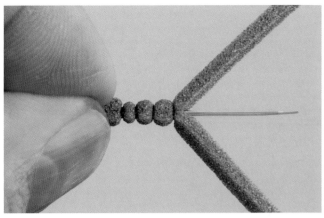

**6.** Repeat steps 4–5 to form 2 more segments, each a little wider than the previous one. After forming the last segment, place 3 half hitches directly over the thread wraps. Trim off the tying thread. Pinch the needle behind the last segment, being careful not to bend or break the tails, and push the foam body off the needle.

**7.** Gently fold the tail fibers out of the way, and cut off the end of the rear segment a short distance behind the thread wraps. Removing this fold in the foam will leave 2 short stubs of foam.

**8.** Pull the tail fibers down between the 2 foam stubs. Preen them into a split-tail configuration. Then carefully trim down the perimeter of the rearmost segment so that it matches the diameter of the adjoining segment. Place a drop of cement at the base of the tail fibers and over the thread wraps used to form the body segments. Do not clip the tail butts.

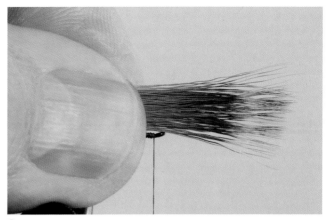

**9.** Mount a hook in the vise. Mount the thread behind the eye, and wrap a thread foundation over the front half of the hook. Position the thread about 2 thread wraps' distance behind the hook eye. Clean and stack a bundle of deer hair as described for the Sparkle Dun, steps 1–4, pp. 27–28. Position the bundle atop the hook so that the tips extend forward of the hanging thread about twice the length of the hook shank.

**10.** Take 3 snug thread wraps to mount the hair. Then slide your left fingers rearward, and take 6 additional tight thread wraps toward the rear of the hook.

**11.** Trim the butt ends of the hair perpendicular to the hook shank just forward of the hook point, and bind them down.

**12.** Wrap the thread to the rear of the hook shank. Coat the thread wraps with cement. Fold the foam extension around the sides of the hook shank, and position it to form a new body segment that is a little wider than the adjoining segment. The butts of the tail material should lie along the hook shank, sandwiched between the foam halves. Secure the foam with 4 snug thread wraps.

**13.** Fold the foam tag ends rearward, and trim away the tail butts. Wrap the thread forward to a point about 4 to 6 thread wraps' distance behind the butt ends of the wing, as shown in this top view. Coat the thread wraps with cement.

**14.** Fold the foam forward, and secure it with 4 snug thread wraps. Trim the tag ends of the foam. Leave a gap about 4 thread wraps wide between the wing butts and the front of the abdomen. Position the thread in the middle of the gap. Use a dubbing needle to coat the thread wraps with cement.

**15.** Preen the deer hair rearward around the hook shank, and take 3 tight thread wraps to secure it.

**16.** Release your fingers from the hairs; they should flare out from the hook shank. Gently preen the hairs to form a 180-degree arc over the top of the shank. Trim away any hairs underneath the shank. Take 2 to 3 tight thread wraps over the previous wraps to lock the hair in place; then tie off and cut the thread. Apply cement to the base of the wing and the wing-mounting wraps.

# Brown Drake Spinner Pattern

## POLY-FOAM SPINNER

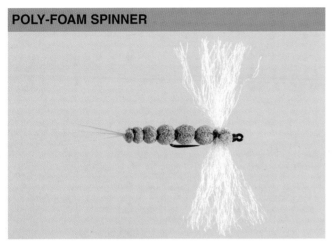

| | |
|---|---|
| **Hook:** | #12-14 TMC 206BL |
| **Thread:** | Brown 6/0 |
| **Tails:** | Tan paintbrush bristles |
| **Body:** | Tan Larva Lace Dry Fly Foam |
| **Cement:** | Dave's Flexament, or similar flexible cement, thinned 1-to-1 with solvent |
| **Wings:** | Tan poly yarn |

Though trout don't ordinarily become selective to spinners during the Brown Drake hatch, we carry this pattern as a backup because trout are unpredictable. Even in early twilight, this flush-floating pattern is difficult to see on the water. Fish it with the same kind of twitches that you would use to fish the dun imitation. To ensure that these long spentwings hold their shape, use a stiff poly yarn such as Spirit River Poly-Bear Fiber. Tying this fly is similar to tying the Buoyant Brown Drake, so the steps here are abbreviated. See pp. 79 and 80 for notes about the foam material and paintbrush bristles used for this pattern.

**1.** Using a 3-inch strip of ⅛ x ³⁄₃₂-inch (3 x 2 mm) closed-cell foam, form an extended foam body and tails as described for the Buoyant Brown Drake, steps 1–8, pp. 80–81. Mount the thread behind the hook eye, and lay a foundation to the rear of the hook shank. Coat the thread wraps on the rear half of the shank with cement. Fold the foam extension around the

sides of the hook shank, and position it to form a new body segment that is a little wider than the adjoining segment. The butts of the tail material should lie along the hook shank, sandwiched between the foam halves. Secure the foam with 4 snug thread wraps.

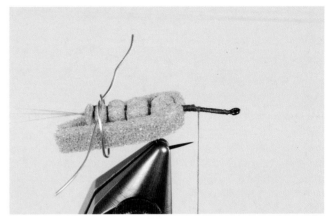

**2.** Fold the tag ends of the foam rearward, and fold a short length of lead wire over them to hold them in place. Trim away the butts of the tail fibers, and secure them, wrapping forward until the hanging thread is 2 to 3 thread wraps' distance ahead of the hook point.

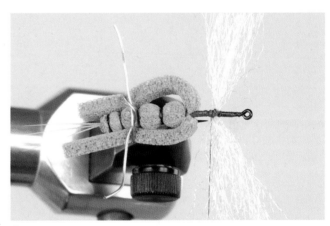

**3.** Clip a 2-inch length of poly yarn for the wings. Mount the yarn crosswise atop the shank using tight crisscross wraps. Once the yarn is secure, take 2 thread wraps directly behind the wings.

**4.** Preen the wings forward, and apply a drop of cement to the thread foundation on the shank behind the wings. Remove the wire from the foam. Fold the foam tags forward and slightly downward. Take 4 thread wraps around the foam directly behind the wing.

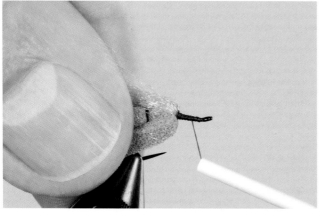

**5.** Draw the wings and foam tags rearward, and wrap the thread forward to a point 4 thread wraps' distance behind the hook eye.

**6.** Fold the foam strips forward under the wings, and secure them with 3 thread wraps.

**7.** Trim the excess foam, bind it down, and finish the head of the fly. Hold the wings vertically, and cut them to about the length of the body.

**8.** Preen the wings into a horizontal position, and place a drop of cement at base of the wings to lock the fibers in place, as shown in this front view.

# Callibaetis

*Callibaetis* are most commonly found in the weedy margins of lakes, but they also inhabit trout streams with slow currents and rooted vegetation—spring creeks, slow-flowing tailwaters, meadow streams, and the inlet and outlet streams of lakes. In the proper kind of moving-water habitat, populations can be very large and the hatches dense. As a rule, these mayflies produce two or three generations a year—in the spring and fall on some waters and in the spring, summer, and fall on others. As a result, *Callibaetis* offer abundant angling opportunities and, in some locations, season-long fishing. As with other multigenerational mayflies, each new generation during the season is a bit smaller than the preceding one. Because of the varying sizes and the range of colors exhibited by these insects, it's important to capture a natural and match your fly to it. Trout see a lot of these mayflies over the course of a season and can grow very selective.

**Other common names:** Speckle-Wing Quill, Speckle-Wing Dun
**Family:** Baetidae
**Genus:** *Callibaetis*
**Emergence:** April to October, midday during the cooler months, early morning or evening during the warmer months
**Spinnerfall:** Midday to evening; generally in the warmest part of the day during the cooler months and in the evening during the warmer months
**Body length:** ¼–½ inch (6–12 mm)
**Hook sizes:** #12-18, with #14-16 most common

**Left:** Callibaetis *nymph: tan to dark tan, or olive to olive-brown*

**Above:** Callibaetis *nymph (underside): lighter shade of top color.*

**Left:** *Male* Callibaetis *dun: gray to olive-brown; gray-brown wings with distinct white venation.*

**Above:** *Male* Callibaetis *dun (underside): lighter shade of the top color, often with an olive tint.*

**Above:** *Female* Callibaetis *spinner: same body color as males; clear wings with mottling on leading edge of forewings.*

**Top right:** *Female* Callibaetis *spinner (underside): lighter shade of top color.*

**Right:** *Male* Callibaetis *spinner: brown to gray on top and a lighter shade of top color on underside; clear wings with little or no mottling on leading edge of forewings; large eyes.*

# Important Fishing Stages

**Nymph** *Callibaetis* nymphs are agile, athletic swimmers that spend most of their time sheltered in the vegetation. Trout sometimes root them out of the cover, but more often the fish prowl the edges of weeds looking for nymphs that have ventured into open water. As the nymphs mature, they grow more active, often swimming upward a short distance and then drifting or darting back to cover. At emergence, they swim to the surface, where the duns hatch. Though you can fish a nymph pattern all season long, the most productive times are just prior to emergence and just as the hatch is beginning, before the trout start taking emergers or duns from the surface. Fish a nymph imitation by itself, using a slow retrieve that keeps the fly just above the weeds. Fishing the nymph pattern as dropper below an indicator or dry fly can also be quite effective: adjust the length of the tippet so that the fly is close to the bottom, then occasionally twitch the fly lightly during the drift to give it lifelike movement.

**Emerger/dun** During cooler weather, the duns spend a lot of time on the surface of the water; as the temperatures warm, they become airborne more quickly. The best dry-fly fishing often occurs in the spring and fall on cold, blustery days, when the duns seem never to leave the water, and the trout have ample opportunity to capture them. But even during the warmer months, the duns offer decent dry-fly fishing, though the hatches tend to be shorter in duration, and the adult flies spend less time on the water.

**Spinner** Mating swarms generally collect above the shoreline. After mating, the spent males usually fall to the ground. The females fly into the foliage, where they stay for about five days, until their eggs are ready to hatch. The females then return to the water to deposit their eggs, after which they fall spent to the surface. Since the clear-winged male spinners typically fall on land, the mottle-winged females are most available to the trout; they are the spinners you should imitate.

# *Callibaetis* Nymph Pattern

## MARABOU NYMPH

| | |
|---|---|
| **Hook:** | #12-16 standard nymph |
| **Weight:** | Lead or nontoxic wire |
| **Thread:** | Brown 8/0 |
| **Tail:** | Natural or dyed-olive pheasant tail-feather barbs |
| **Rib:** | Fine copper wire |
| **Abdomen:** | Olive-brown marabou |
| **Wing case:** | Pheasant tail-feather barbs |
| **Thorax:** | Olive-brown dubbing |
| **Legs:** | Tips of pheasant tail barbs from wing case |

It's important when dressing this fly to use only a few wraps of lead or non-toxic wire. The pattern should be kept light for a couple of reasons. First, the fly is typically fished in very slow water, and too much weight will cause it to sink in the weeds or catch on the bottom. Second, when the natural insect swims up toward the surface, it often pauses periodically for a few seconds, and during this time it slowly sinks a few inches; if the fly is too heavy, it will sink too fast, behaving unnaturally. Shortly before a hatch and during emergence, use a 3- to 6-inch strip retrieve with short pauses, and fish the fly all the way to the surface. At any other time, fish the pattern close to the weeds below an indicator, with an occasional twitch to give it some life.

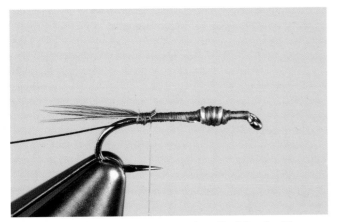

**1.** Form an underbody of wire as described for the Pheasant Tail Nymph, steps 1–2, p. 21. The front of the underbody should be about ¼ of a shank length behind the eye. Mount the thread behind the eye, and secure the wire wraps. Mount a length of ribbing wire at the midpoint of the shank, and wrap it to the tail-mounting point, binding the wire to the near side of the shank as you wrap. Align the tips of 3 pheasant tail barbs, and mount them atop the shank to form tails one hook gap in length.

**2.** Align the tips of 4 marabou barbs, and mount them atop the shank at the tail-mounting point. Advance the tying thread to the midpoint of the shank. Gather the marabou barbs into a bundle. Clip the butt ends in a pair of hackle pliers, and gently twist the bundle clockwise (when viewed from above) to form a chenille-like strand, as shown in this top view.

**3.** Wrap the marabou to the midpoint of the shank. Tie off the material, and clip the excess. Counterwrap the ribbing wire—that is, wrap it in the opposite direction from the one you wrapped the marabou in—in an open spiral of 3 or 4 turns over the abdomen. Secure the rib in front of the abdomen, and clip the excess. Position the thread at the front of the abdomen.

**4.** Align the tips of 8 to 10 pheasant tail barbs; mount the bundle directly ahead of the abdomen so that the barbs' tips extend to the end of the tail. Clip the excess, and position the thread over the rearmost mounting wrap.

**5.** Dub a thorax slightly larger in diameter than the abdomen, stopping 4 to 5 thread wraps' distance behind the hook eye. Fold the pheasant tail barbs smoothly and uniformly over the top of the thorax, and secure them at the front of the thorax.

**6.** Fold half of the pheasant tail fibers rearward along the near side of the shank, and half along the far side of the shank. Pinch them in this position with your left fingers, and take 4 to 5 thread wraps over the folds in the barbs to slant the fibers rearward, as shown here from the top. Then finish the head of the fly.

# *Callibaetis* **Emerger Pattern**

## *CALLIBAETIS* EMERGING DUN

*Originator: Wayne Luallen*

| | |
|---|---|
| **Hook:** | #14-16 standard dry-fly |
| **Thread:** | Olive 8/0 |
| **Tail:** | Mallard flank-feather barbs topped with light brown Antron yarn fibers |
| **Rib:** | Dark brown thread |
| **Abdomen:** | Olive-tan dubbing |
| **Wing:** | Light dun CDC feather topped with mallard flank feather |
| **Thorax:** | Gray dubbing |
| **Head:** | Light olive dubbing |

This elegant design combines a slim, low-riding body with a pronounced CDC loop wing topped with mallard that offers good flotation and markings strikingly like those of the natural insect. If you leave the fly untreated, or carefully apply powdered floatant to the wing only, the abdomen will absorb water and sink below the surface film. But the pattern sometimes works better if the entire fly is treated with powder floatant to ride flush on the surface. The wing on this fly can be a little fragile, particularly if a trout is hooked in the toothy part of the jaw. But for us, its effectiveness outweighs this fairly minor drawback. A tattered fly still catches fish; it's just harder to keep afloat.

**1.** Mount the thread behind the hook eye, and lay a thread foundation to the tailing point. Align the tips of 3 mallard flank-feather barbs, and mount them atop the shank to make a tail one hook gap in length. Gently pull them apart with your fingers to form 3 distinct tails. Return the thread to rearmost thread wrap, and mount a very sparse bundle of Antron that reaches halfway to the end of the tail. Bind and clip down the excess materials. Return the thread to the tailing point.

**2.** Mount a 6-inch length of ribbing thread. Then dub a slender, slightly tapered abdomen to the midpoint of the shank. Rib the body in an open spiral of 6 to 8 wraps. Tie off the ribbing, and clip the excess. Dub a thorax halfway to the hook eye. Position the thread at the front edge of the thorax.

**3.** Preen back the lower barbs on a mallard flank feather, leaving a tip section containing 10 to 15 barbs. Place the feather atop the shank, with the tip forward and concave side facing up. Take one wrap over the feather stem exposed at the gap in the barbs, as shown in this top view.

**4.** With your right hand, put gentle downward pressure on the tying thread. With your left hand, pull the butt end of the mallard feather, sliding it beneath the thread wrap until only about 1/4 of the feather tip extends beyond the thread wrap. This "slide mount" helps keep the feather barbs smooth and aligned.

**5.** Bind down the tip of the mallard feather, and return the thread to the rearmost thread wrap. Preen back the short, scraggly barbs from the base of a CDC feather until you reach a point on the stem where the barbs are reasonably uniform in length. Position the feather atop the shank, just as you did with the mallard flank in step 3. Then use the same "slide mount" shown in steps 3–4 to mount the CDC feather directly atop the mallard flank. Bind down and clip the excess. Position the thread at the rearmost wrap.

**6.** Dub the thorax halfway to the hook eye.

**7.** Position a dubbing needle crosswise atop the thorax. Fold both feathers over the needle to create a loop.

**8.** Hold the loop in position with your left fingers, and secure the feathers directly in front of the thorax. Bind and clip the excess.

**9.** Dub a short, tapered head, and finish the fly. Shown here is a top view of the wing. Note that the feather barbs do not form a narrow band over the thorax, but spread a little to the sides of the shank.

# *Callibaetis* Dun Pattern

## SPECKLED-WING PARACHUTE

*Originator: Pret Frazier*

| | |
|---|---|
| **Hook:** | #12-18 standard dry-fly |
| **Thread:** | Olive 8/0 |
| **Tail:** | Light dun Microfibetts |
| **Body:** | Light olive-brown |
| **Wing post:** | Mallard flank feather |
| **Hackle:** | Medium dun dry-fly |

Mallard flank feathers do a highly credible job of imitating the mottled gray wings of the *Callibaetis* dun, and on this parachute pattern, they are easily seen on the water by both the fish and the angler. The wing on this pattern is deliberately mounted to be overly long; trimming away the very thin barb tips once the fly is complete gives the wing a clearer, more distinct profile. Some tiers prefer this pattern with split tails, which you can form as described for the *Baetis* CDC Compara-dun, steps 3–7, pp. 11–12, or you can dress the flared tail shown below and trim away the center tail fibers. Dressed with a lighter body color and light dun hackle, this fly makes a very good spinner pattern.

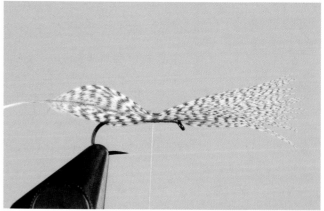

**1.** Mount the thread behind the hook eye, and wrap a foundation to the midpoint of the shank. Return the thread to a point ⅓ of a shank length behind the eye. Select a well-marked mallard flank feather with a squared-off tip. Smooth the barbs toward the tip. The central section of the barbs will have their tips aligned; the barbs near the edges of the feather are shorter. Strip away these short barbs so that all the remaining barbs are aligned at the tips. Mount the feather atop the hook so that the barb tips extend 2 shank lengths beyond the mounting wraps.

**2.** Clip away the butt of the feather, and secure with additional thread wraps. Lift the wing fibers vertically. Take several tight thread wraps against the front base of the bundle, building a bump of thread that will keep the wing fibers upright. Build a foundation for the hackle with 8 to 10 thread wraps as described for the Parachute BWO, step 2, p. 15. Prepare and mount a parachute hackle as shown in the same pattern, steps 3–4, p. 16.

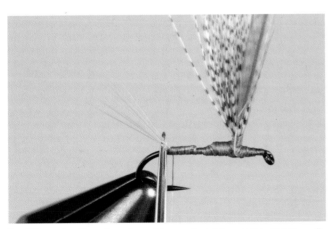

**3.** Wrap the thread to the tailing point. Mount 6 to 8 Microfibetts as a bundle to form a tail one shank length long. Clip and bind the excess. To flare the tail fibers, push your thumbnail or a dubbing needle against the base of the tail from underneath. Once the fibers are flared out, you can adjust them into a symmetrical fan shape by positioning them with your fingers.

**4.** Use a sparsely dubbed thread to form a slender, tapered body, stopping 4 to 5 thread wraps' distance behind the eye.

**5.** Wrap and tie off the hackle as described for the Parachute BWO, steps 7–10, pp. 16–17. Finish the head of the fly. Use your fingers to pull the wing fibers toward the hook eye and the hook bend, fanning them into a broader silhouette.

**6.** Trim the wing to be one shank length in height above the topmost hackle wrap. Round the tips of the wings for a more realistic profile.

# *Callibaetis* Spinner Pattern

**CALLIBAETIS GATHERED-HACKLE SPINNER**

| | |
|---|---|
| **Hook:** | #12-18 standard dry-fly |
| **Thread:** | Tan 8/0 |
| **Tail:** | Light dun Microfibetts |
| **Rib (optional):** | Light brown thread |
| **Body:** | Light gray dubbing |
| **Hackle:** | Grizzly dry-fly, one size larger than normally used for the hook size |

This clean, uncluttered design is simple to tie and makes a highly effective imitation of the natural insect; the grizzly hackle reproduces the glassy, mottled wings of the female spinner. Forcing the hackle barbs beneath the shank to the horizontal position increases the density of the wing, improves flotation, and places the body of the fly flush against the surface. Leaving the hackle barbs above the shank intact makes this pattern more visible on the water than other flush-floating styles. Keep the body very slender.

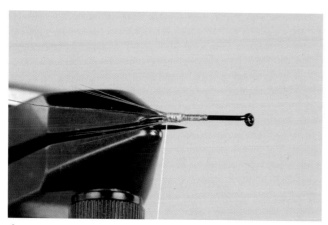

**1.** Mount the thread at midshank, and wrap to the tailing point. Use 3 Microfibetts per side to create split tails that are one hook length long, as described for the *Baetis* CDC Compara-dun, steps 3-7, pp. 11–12, and shown in this top view. Bind and clip the excess. If dressing the optional rib, mount a 6-inch length of ribbing thread over the rearmost tailing wrap.

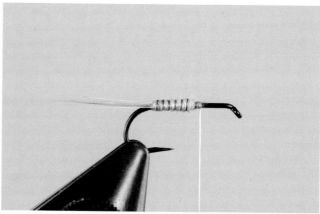

**2.** Using a sparsely dubbed thread, wrap a slim, slightly tapered abdomen to the midpoint of the shank. If dressing the rib, spiral the ribbing thread in 5 to 6 wraps over the abdomen. Tie off and clip the ribbing. Position the thread 3 thread wraps' distance in front of the abdomen.

**3.** Prepare and mount a hackle as described for the Little Green Drake, steps 4–6, p. 42. Position the thread 6 to 7 thread wraps' distance behind the hook eye. Wrap the hackle forward, taking 6 to 8 wraps of the feather, and secure the feather tip as explained in steps 7–10, pp. 42–43. Clip the excess.

**4.** If you tie on a rotary vise, spin the jaws to turn the fly upside down. On a conventional vise, remove the fly from the jaws and remount it upside down. Using your fingers, create a gap in the hackle collar by preening the barbs to the sides of the shank. Wrap the thread through the gap, over the shank, and down the far side of the hook behind the hackle. Make this wrap tight; the aim is to force the hackle barbs to the horizontal position.

**5.** Take one wrap of thread around the hook shank behind the rearmost hackle wrap. Bring the thread up behind the hackle, through the gap again, and down the far side of the shank ahead of the hackle, completing a crisscross wrap through the gap in the hackle.

**6.** Take a wrap of thread around the shank in front of the hackle. If the hackle barbs are not yet quite horizontal, repeat the crisscross wrap until the barbs are forced into a horizontal position. Then, with a sparsely dubbed thread, repeat the crisscross wrap one or two more times to cover the bare thread wraps and bring the thorax to size.

**7.** When the hackle is properly positioned, with the lowermost barbs extending horizontally outward from the hook shank, as shown in this front view, take a wrap or two of dubbing toward the hook eye, and finish the fly.

# Gray Drake

The Gray Drake is rather unusual for a mayfly in that the nymphs crawl out of the water to hatch, behaving more like stoneflies than like other mayfly species. Consequently, the angler faces the curious circumstance of a hatch in which emergers and duns are virtually insignificant, since these forms are not routinely available to the trout, which feed mainly on the nymphs and spinners. Like the Brown Drake, the Gray Drake has specialized living requirements; ideal habitat is a stream with moderate to slow currents, flowing between root-lined banks over a weedy, silty bottom. Such waters are scattered throughout the West, but as regional trout streams go, they are the exception rather than the rule. Typically, these waters are not easy to fish because of their mucky bottoms and brushy banks. Gray Drake populations vary with each stream. Some hatches are too sparse to be considered fishable; others are so dense that the profusion of spinners becomes something of a liability. Getting your fly noticed among all the naturals can be difficult, and such hatches may be fishable only at the beginning of the emergence. But the Gray Drake hatch is a noteworthy event on some waters, and when these large mayfly spinners are falling spent in decent numbers, they will bring most of the trout in the stream to the surface.

| | |
|---|---|
| **Other common name:** | Black Drake |
| **Family:** | Siphlonuridae |
| **Genus:** | *Siphlonurus* |
| **Emergence:** | Mid-May to mid-July in the Pacific region, August and September in the Rocky Mountain region, sporadically any time of the day or night |
| **Spinnerfall:** | Midmorning to late evening |
| **Body length:** | ⅜–¾ inch (10–20 mm) |
| **Hook sizes:** | #8-12 |

**Left:** *Gray Drake nymph: shades of brown to gray.*

**Above:** *Gray Drake nymph (underside): lighter shade of top color.*

**Right:** *Gray Drake dun: light gray to dark gray; light gray wings.*

**Left:** *Female Gray Drake spinner: dark gray to almost black, often with a reddish tint and usually with distinct banding on the abdomen; clear wings.*

**Above:** *Female Gray Drake spinner (underside): lighter shade of top color.*

# Important Fishing Stages

**Nymph** Gray Drake nymphs are excellent swimmers; by flipping their hair-fringed tails, they propel themselves rapidly in short bursts. When the nymphs venture into open water, they are readily taken by trout. Most of the time, however, the nymphs remain sheltered or hidden in the vegetation or silt, and trout occasionally hunt them out and dislodge them. As emergence nears, the mature nymphs move to shallows near the banks, and they are at their most vulnerable to predation during this migration. You can fish a nymph pattern any time of the day during the several weeks leading up to the hatch, but the best time is a week or two before emergence begins. Fish the nymph pattern with short strips close to the bottom or along weed-lined banks. On streams with silty beds, let the fly sink to the bottom and rest in the silt for a few moments. Then give the rod tip a quick upward twitch to lift the fly off the bottom, and use a quick strip to retrieve 4–10 inches of line. Stop and let the fly settle to the bottom; repeat this technique throughout the drift.

**Emerger/dun** Because the nymphs crawl out of the water to emerge, emergers and duns normally are not fishable phases of the life cycle. However, on streams where submerged weeds reach to the surface in flowing water, some nymphs may crawl out on this vegetation to hatch, and some duns may end up drifting down the current. These are large mayflies, and it doesn't take many of them on the surface to get the trout interested. If you are lucky enough to find duns on the water, you can use the spinner pattern described in this section, though the Parachute Adams, also in this section, is a better match.

**Spinner** The dark-bodied spinners are the reason this mayfly is often called a Black Drake. After the duns emerge, they molt into spinners in a day or two, and then mate in swarms over or alongside the stream. Shortly after mating, the females dip to the surface, release their eggs, and fall to the water. For a short time after landing on the water, many of the drifting spinners hold their wings upright before lowering them to the surface; this gives some anglers the false impression that there are duns on the water and a hatch is occurring. Once a spinner falls to the surface, it seldom moves, so fish spinner imitations drag-free, without adding any twitches to the fly.

# Gray Drake Nymph Pattern

**GRAY DRAKE SHELLBACK**

| Hook: | #12-14 2X–3XL nymph |
|---|---|
| **Weight:** | Lead or nontoxic wire |
| **Thread:** | Brown or gray 6/0 |
| **Tail:** | Dark brown marabou |
| **Shellback:** | Brown Thin Skin or Fino Skin strip |
| **Rib:** | Fine copper wire |
| **Abdomen:** | Dark brown marabou |
| **Thorax:** | Peacock herl |
| **Legs:** | Pheasant tail-feather barbs |

The ribbed marabou abdomen on this pattern nicely suggests the conspicuous gills on this nymph and, along with the marabou tail, gives the fly a mobility that imitates the natural insect, which is an active swimmer. While marabou bodies can be somewhat fragile, the shellback protects the abdomen and makes for a durable fly. When tying the body, look for marabou feathers with long, very fluffy barbs.

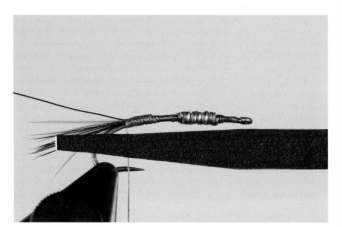

**1.** Weight the thorax area of the hook as described for the Pheasant Tail Nymph, steps 1–2, p. 21. Wrap a thread foundation to the rear of the shank. Align the tips of 8 to 10 marabou barbs, and mount them as a bundle atop the shank to form a tail about as long as the hook shank. Clip the excess. Mount a 3-inch length of ribbing wire, and position the thread at the tail-mounting point. Cut a strip of Thin Skin one hook gap in width. Trim it to a taper so that the end is about ⅓ of a hook gap in width.

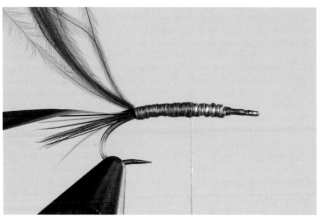

**2.** Mount the tapered end of the strip shiny side up. Bind down the excess, and return the thread to the tail-mounting point. Align the tips of 6 to 10 marabou barbs. Trim away about ¼ inch from the tip end of the bundle, and mount the tips. Wrap the tying thread forward to the midpoint of the shank, creating a level underbody as you go.

**3.** Gather the barbs smoothly together, and clip the butt ends in a pair of hackle pliers. Gently twist the bundle of barbs clockwise (as seen from above) to form a chenille-like strand. Wrap the barbs to the midpoint of the shank, and secure the strand.

**4.** Clip the excess marabou, and position the tying thread directly in front of the abdomen. Draw the Thin Skin smoothly over the top of the abdomen. Secure it in front of the abdomen with the thread, but don't clip the excess. Counter-rib the abdomen by wrapping in the opposite direction from the one in which you wrapped the marabou, taking 5 to 6 evenly spaced turns. Secure the ribbing wire in front of the abdomen, and clip the excess.

**5.** Fold the Thin Skin rearward. Wrap the tying thread rearward, over the fold in the Thin Skin, until you reach the front of the abdomen. Mount 3 to 5 peacock herls and a length of scrap tying thread, and form the thorax as described for the Pheasant Tail Nymph, steps 7–9, pp. 22–23, stopping 5 to 7 thread wraps' distance behind the hook eye. Tie off the herl and clip the excess. Draw the Thin Skin over the thorax, and secure it directly in front of the thorax.

**6.** Clip and bind down the excess Thin Skin. Align the tips of 3 pheasant tail barbs, and mount them against the near side of the shank to form legs one hook gap in length. Repeat the process to form legs on the far side, as shown in this top view. Trim the excess, and finish the head.

# Gray Drake Dun Pattern

**PARACHUTE ADAMS**

| | |
|---|---|
| **Hook:** | #10-12 standard dry-fly |
| **Thread:** | Gray 6/0 |
| **Tail:** | Mixed grizzly and brown dry-fly hackle barbs |
| **Body:** | Gray dubbing |
| **Wing post:** | White calf tail |
| **Hackle:** | 1 brown and 1 grizzly dry-fly |

Though we rarely dress dun patterns specifically for the Gray Drake hatch, we are seldom without a Parachute Adams, which, in the sizes noted, makes a workable imitation of the dun. The spinner pattern below can also be used, but the Parachute Adams floats better, and the white wing post makes it much more visible.

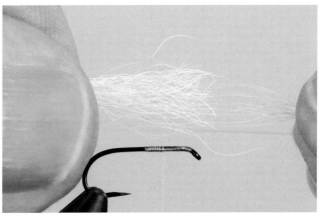

**1.** Mount the thread behind the hook eye. Wrap a thread foundation to the midpoint of the shank, and position the thread $1/3$ of a shank length behind the eye. Clip a bundle of calftail hair, and clean out the shorter hairs as shown in the instructions for the Sparkle Dun, step 2, p. 27. Then pinch the base of the hair bundle, and pull out the longer hairs, as shown, so that the tips of the hairs remaining in the bundle are approximately aligned.

**2.** Mount the hair atop the shank, tips forward. The bundle of hair ahead of the mounting wraps should be one shank length long. Clip the butts, and bind the excess. Lift the wing fibers vertically. Take several tight thread wraps against the front base of the bundle, building a bump of thread that will keep the wing fibers upright. Build a foundation for the hackle with 8 to 10 thread wraps as described for the Parachute BWO, step 2, p. 15. Position the thread behind the wing post.

**3.** Prepare the hackle feathers as described for the Parachute BWO, step 3, p. 16. Stack the two feathers so that the lowermost barbs are aligned and the curvatures match. It doesn't matter which feather is on top. Mount them just as you would a single feather, as explained in steps 3–4, p. 16.

**4.** Prop the hackle feathers against the front of the wing post to keep them out of the tying field. Wrap the thread to the tailing point. Align the tips of 4 or 5 brown hackle barbs; then align the tips of 4 or 5 grizzly barbs. Next, align the tips of all the barbs, and combine them into a single bundle. Mount it to form a tail one shank length long.

**5.** Using a sparsely dubbed thread, form a slightly tapered body, stopping 3 to 4 thread wraps' distance behind the eye.

**6.** Wrap and tie off the top feather as described for the Parachute BWO, steps 7–10, pp. 16–17. Take the first wrap of this feather just below the top of the thread foundation on the wing post.

**7.** Spiral the bare stem of the second feather up the post, wiggling it up and down as you go so that it slides between the barbs of the wrapped feather. Take the first wrap of this second feather at the very top of the thread foundation on the post.

**8.** Wrap the feather down the post, again wiggling it up and down to slide the wraps between the barbs of the first feather. Take about as many wraps of the second feather as you did the first. Tie off the second feather just as you did the first. Trim the excess, and finish the head.

# Gray Drake Spinner Pattern

**CLIPPED-HACKLE SPINNER**

| | |
|---|---|
| **Hook:** | #10-12 standard dry-fly |
| **Thread:** | Black 6/0 or 8/0 |
| **Tail:** | Dark dun Microfibetts |
| **Rib:** | Maroon 3/0 thread |
| **Body:** | Dark gray dubbing |
| **Hackle:** | Light dun dry-fly, one size larger than normally used for the hook size |

The upright hackle fibers make this pattern relatively easy to see on the water, which is a big advantage during a heavy spinnerfall, when it can be difficult to spot your fly among so many naturals on the surface. This fly is tied in a fashion very similar to that shown for the *Callibaetis* Gathered-Hackle Spinner pattern, pp. 92–93, so the instructions here are abbreviated. In fact, if you wish, you can use the instructions for that pattern, gathering the hackle with a dubbed thread rather than trimming it as described here. Trimming is faster, but gathering the hackle puts more barbs into contact with the surface film. Either way works.

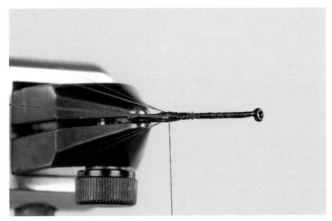

**1.** Mount the thread behind the eye, and wrap a thread foundation to the rear of the shank. Use 3 Microfibetts per side to create split tails that are as long as the hook shank, as shown here from the top and described for the *Baetis* CDC Comparadun, steps 3–7, pp. 11–12. Bind and clip the excess tail material. Mount a 6-inch length of ribbing thread over the rearmost tailing wrap.

**2.** Using a sparsely dubbed thread, wrap a slim, slightly tapered abdomen to the midpoint of the shank. Spiral the ribbing thread in 5 to 6 wraps over the abdomen. Tie off and clip the ribbing. Position the thread directly in front of the abdomen.

**3.** Prepare and mount the hackle as described for the Little Green Drake, steps 4–6, p. 42. Position the thread 6 to 7 thread wraps' distance behind the hook eye. Wrap the hackle forward, taking 6 to 8 wraps of the feather, and secure the feather tip as explained in steps 7–10, pp. 42–43.

**4.** Clip and bind the feather tip, and finish the head of the fly. Trim the hackle barbs on the underside of the hook to be flush with the shank, as shown in this front view.

# *Hexagenia*

*Hexagenia* mayflies thrive only in streams that meet very particular requirements. Because the nymphs construct U-shaped burrows in the streambed, they need a muddy bottom that is soft enough to dig into but firm enough to keep the burrows from collapsing. This type of bottom is most often found in slow-moving streams with substrate that is fairly stable and not flushed out by annual spring runoff. Though by no means commonplace, such streams occur in places throughout the West, and those containing good Hex populations are generally well known. The enormous size of these insects and the feeding frenzy they can provoke among the trout make the Hex a super hatch in those localities where it is found. Emergence and egg laying ordinarily begin around twilight and continue into the night, though on cloudy days they may occur earlier. It's worth checking local regulations, since some states prohibit fishing after dark.

| | |
|---|---|
| **Other common name:** | Big Yellow May |
| **Family:** | Ephemeridae |
| **Genus:** | *Hexagenia* |
| **Emergence:** | June to August, twilight into the night |
| **Spinnerfall:** | Twilight into the night |
| **Body length:** | ¾–1½ inches (16–37 mm) |
| **Hook sizes:** | #8, 2XL–3XL |

**Left:** Hexagenia *nymph: yellow-brown.*

**Above:** Hexagenia *nymph (underside): lighter shade of top color.*

**Above:** Hexagenia *dun: light tan to bright yellow with distinct dark markings on top; lighter shade of top color on underside; yellow wings.*

**Left:** Hexagenia *spinner: light tan or yellow-brown to red-brown with dark markings on top; lighter shade of top color on underside; clear wings with a yellow tint.*

# Important Fishing Stages

**Nymph** *Hexagenia* nymphs normally remain concealed during the daylight hours and leave their burrows only at night to feed. When the nymphs mature, they leave the streambed and swim to the surface, where the duns emerge. The nymphs are strong, active swimmers, propelling themselves with rapid up-and-down undulations of their bodies, pausing periodically. This behavior makes it imperative to fish a pattern with a lot of flex and mobility in a fashion that mimics the body movement of the natural. The weeks just prior to the hatch and during the hatch cycle are the best times to use a nymph imitation. Before the hatch starts, fish the fly close to the bottom; once emergence has begun, let the fly sink to the bottom, and then fish it back to the surface in short strips with pauses between them.

**Emerger/dun** The transformation from nymph to dun ordinarily occurs quite quickly. As in all hatches, there are laggards and stillborns, and trout will certainly take an emerging fly. But the transitional phase is so short that we have never found the need to imitate the emerger specifically, though some anglers do. The vast majority of the time, we find that a dun pattern suffices. After a Hex hatches, it takes anywhere from a few seconds to a minute before it flies off the water, and during this time, the dun often moves about, fluttering its wings. Trout key in on this movement, which should be imitated when you fish. After making a cast, strip in any slack and let the fly sit for a short time; give it a slight twitch and strip in the slack again. Repeat this twitch and pause throughout the drift. As with other night hatches, you must fish more by sound and feel than by sight. If you minimize slack, you may feel the take, and the trout may hook itself. But it's best to give a short, light hook set whenever you hear a rise in the vicinity of your fly or at the slightest disturbance of your line. If it proves to be a false alarm, let the fly resume its drift.

**Spinner** After the duns leave the water, they molt into spinners in about two to three days. The spinners gather in swarms after dark, and mating takes place. Afterward, the females return to streamside foliage for a short time to extrude their eggs, then fly back to the water, land on the surface, and release the eggs. Spinners are frequently on the water at the same time as the duns, and when they first land on the water, their wings are upright for a short time before falling spent on the surface. Egg-laying females typically wiggle their abdomens to release their eggs, and the same strip-and-pause technique used for a dun pattern imitates the behavior of the spinner. Once the spinners have finished laying their eggs, their wings droop to the surface, and they stop moving their bodies. Trout do key in on these immobile, spent spinners, but even so, an occasional twitch of the fly is often more effective than leaving it motionless. As a rule, a dun pattern can be used to imitate a spinner, but it's worth carrying a spent-wing pattern for those times when the trout are selective to spent spinners.

# Hex Nymph Pattern

**HEX STRIP NYMPH**

*Originator: Gary Borger*

| | |
|---|---|
| **Hook:** | #6-8 standard nymph |
| **Weight:** | Lead or nontoxic wire |
| **Thread:** | Tan 6/0 |
| **Abdomen:** | Gold rabbit strip |
| **Thorax:** | Gold Hare-Tron dubbing |
| **Wing case:** | Peacock herl |
| **Legs:** | Fibers from thorax |

Because Hex nymphs are so vigorous, the static, stiff-bodied nymph imitations that prevail in fly fishing work poorly during this hatch. A design and materials that enable the fly to mimic the sinuous movements of the natural insect are substantially more effective. This pattern has both. On a strip-and-pause retrieve, the weighted thorax creates a jigging effect that gives the long, flexible hair-strip abdomen an undulating, swimming motion. This fly has the added benefit of being very easy and quick to tie.

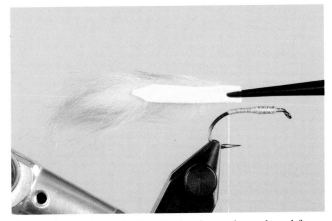

**1.** Mount the thread behind the hook eye, lay a thread foundation to the tailing point, and return the thread to the midpoint of the shank. Weight the thorax area as described for the Pheasant Tail Nymph, steps 1–2, p. 21. Position the thread at the tail-mounting point. Cut a strip of tanned rabbit hide that, from the rear hair tips to the front edge of the strip, is about twice the overall hook length. Remove the fur from the front edge of the strip, leaving about ¼ inch of bare hide. Trim the back end of the strip to a point.

**2.** Mount the fur strip atop the hook, securing it by the bare tab of hide. Bind the whole tab tightly to the shank. Return the thread to the rearmost tail-mounting wraps.

**3.** Dub the front portion of the abdomen over the rear ⅓ of the hook shank. Align the tips of 8 to 10 peacock herls; trim off about ¼ inch from the tip of the bundle. Mount the bundle atop the shank directly ahead of the abdomen. Bind and clip the excess. Position the thread at the rearmost wrap used to mount the herl.

**4.** Dub the thorax to within 4 thread wraps' distance of the hook eye. Draw the herl in a flat band smoothly over the top of the thorax. Secure it tightly directly in front of the thorax.

**5.** Clip the excess wing-case material, and finish the head of the fly. Use a dubbing needle or teaser to pick out some fibers on each side of the thorax to suggest legs.

# Hex Dun Pattern

**BUOYANT HEX DUN**

| Hook: | #12-14 TMC 206BL |
|---|---|
| Thread: | Brown 6/0 |
| Tails: | Tan paintbrush bristles |
| Body: | Yellow Larva Lace Dry Fly Foam |
| Cement: | Dave's Flexament, or similar flexible cement, thinned 1-to-1 with solvent |
| Head/wing: | Bleached deer hair |

This extended foam-body design is particularly well suited to imitations, like this Hex dun, that are fished in very low-light or pitch-black conditions. The exceptional buoyancy and durability of the pattern minimize the need to fuss with floatant or replace damaged flies in the darkness. Except for the colors of the materials, this fly is identical to the Buoyant Brown Drake and is tied according to the instructions for that fly on pp. 79–82, using a 4-inch strip of ⅛ x ³⁄₃₂-inch (3 x 2 mm) foam. See pp. 79–80 for notes about the foam material and paintbrush bristles used for this pattern.

Front view of finished fly.

# Hex Spinner Pattern

## BUOYANT HEX SPINNER

| Hook: | #12-14 TMC 206BL |
|---|---|
| Thread: | Brown 6/0 |
| Tails: | Tan paintbrush bristles |
| Body: | Yellow Larva Lace Dry Fly Foam |
| Cement: | Dave's Flexament, or similar flexible cement, thinned 1-to-1 with solvent |
| Spentwing: | Pearl Krystal Flash |
| Head/wing: | Bleached deer hair |

We no longer use a standard spent-wing pattern for the Hex hatch, since this version of the Hex dun has proven just as effective at catching trout. Moreover, it floats much better and is far easier to see in failing light than a traditional spent-wing design. The Krystal Flash wing on the fly lies flush against the surface and creates a footprint in the film very much like the spent wings of the natural. Except for the colors of the materials and the addition of Krystal Flash, this fly is identical to the Buoyant Brown Drake, pp. 79–82, so the following instructions are abbreviated. See pp. 79–80 for notes about the foam material and paintbrush bristles used for this pattern.

**1.** Using a 4-inch strip of ⅛ x ³⁄₃₂-inch (3 x 2 mm) foam, form and mount the body extension and winging hair as described in the instructions for the Buoyant Brown Drake, steps 1–14, pp. 80–82. Carefully invert the hook in the vise jaws so that the underside of the shank faces upward, as shown here. Position the tying thread directly in front of the abdomen. Clip a bundle of 6 to 8 strands of Krystal Flash about 4 inches long. Position them directly in front of the abdomen, and mount them with tight crisscross wraps of thread. Don't trim the Krystal Flash yet.

**2.** Return the hook to the normal upright position in the vise. Position the thread directly ahead of the Krystal Flash wings, and form the deer-hair wing as shown in the instructions for the Buoyant Brown Drake, steps 15–16, p. 82. When trimming away the deer hair, gather the Krystal Flash strands in your left hand, and draw them upward toward the rear of the fly to keep them out of the way, as shown here.

**3.** When the deer-hair wing is complete, trim the Krystal Flash at a point just beyond the tips of the deer hair, as shown in this bottom view.

# Mahogany Dun

Mahogany duns can be found in a great many—indeed, most—Western streams. This category of mayflies actually contains more than one species, but they are sufficiently similar in appearance and habits that they can be grouped together and imitated with the same patterns. Depending on the species, however, hatches may occur in the spring or fall. Fall hatches usually provide the best fishing; spring runoff can put rivers out of shape during the early season emergence, and the relative scarcity of other hatches in the fall makes the Mahogany Dun emergence stand out as significant. The populations of these mayflies vary widely from stream to stream, but those waters with moderate to slow flows and weeds, roots, and plant debris at the margins often hold the largest numbers of these insects. During most hatches, there are not a great many duns on the water at the same time, but they typically spend so much time drifting on the surface that the trout gather to feed on them, providing some fine dry-fly fishing.

| | |
|---|---|
| **Other common names:** | Blue Dun, Blue Quill, Iron Dun |
| **Family:** | Leptophlebiidae |
| **Genus:** | *Paraleptophlebia* |
| **Emergence:** | April–May and September–October, midmorning to midafternoon |
| **Spinnerfall:** | Midafternoon to early evening |
| **Body length:** | ¼–½ inch (6–12 mm) |
| **Hook sizes:** | #12-18, with #14-16 most common |

**Left:** *Mahogany Dun nymph: olive-brown to dark brown.*

**Above:** *Mahogany Dun nymph (underside): lighter shade of top color.*

**Right:** *Mahogany Dun nymph* (bicornuta): *the* bicornuta *species, which hatches in the fall, is similar in color to other species but is larger and distinguished by pronounced mandibles.*

**Above:** *Male Mahogany Dun: shades of brown, often with center of abdomen a lighter shade; light gray wings.*

**Top left:** *Female Mahogany Dun: reddish brown on top, with lighter shade of top color on underside; light gray wings.*

**Left:** *Female Mahogany Dun spinner (shown): reddish brown to purplish brown. Male spinner (not shown): brown with the middle of abdomen white to almost clear. Clear wings on both sexes.*

# Important Fishing Stages

**Nymph** Mahogany Dun nymphs can be found in most current flows, though they are poor swimmers and if swept away by the current will drift some distance before finding their way to the bottom again. As they mature, the nymphs migrate to slower waters with weedy, debris-strewn bottoms—tailouts of pools, shallow margins of the stream, and eddies—where they will hatch. At emergence, the nymphs either crawl out of the water to hatch or swim to the surface, where the dun sheds its nymphal shuck in or just below the surface film. Two or three weeks before the hatch, when the nymphs living in faster currents are moving to calmer waters, fish a nymph pattern close to the bottom below riffles and in runs. When emergence occurs, fish the nymph imitation in the slower waters where the duns are hatching; suspend the nymph below a dry fly or indicator, or use a floating nymph pattern.

**Emerger/dun** These mayflies generally do not hatch in profuse numbers. There may be only a scattering of duns on the surface at any one time—it can be easy to overlook them—and the surface is seldom pocked with rises that would indicate a hatch. In the fall, always take a few moments to search edge waters, tailouts, and eddies for the occasional drifting dun or rise. But the amount of time the duns spend on the water gives the fish plenty of opportunity to take them, and some trout will take up a feeding position in the calmer waters where the duns are found. These fish can be very selective and wary. To catch them, the angler often needs flush-floating flies of the approximate size and color of the natural; long, fine tippets; stealth; and patience.

**Spinner** The males fall spent to the water after mating, and they are joined by the females once egg laying has taken place. The spinnerfall generally occurs in the waters below riffles or in runs. The spent spinners often land on the water with their wings raised and eventually lower them to the surface. The upright wings make the flies a little easier to see on the water and can also make a dun pattern effective during a spinnerfall.

# Mahogany Dun Nymph Pattern

**PHEASANT TAIL NYMPH**

*Originator: Frank Sawyer/Al Troth*

| | |
|---|---|
| **Hook:** | #12-16 1XL–2XL nymph |
| **Weight:** | Lead or nontoxic wire |
| **Thread:** | Brown 6/0 or 8/0 |
| **Tail:** | Pheasant tail-feather barbs |
| **Rib:** | Fine copper wire |
| **Abdomen:** | Pheasant tail-feather barbs |
| **Wing case:** | Pheasant tail-feather barbs |
| **Thorax:** | Peacock herl |
| **Legs:** | Pheasant tail-feather barbs |

That the Pheasant Tail Nymph is considered one of the great all-purpose patterns is no accident. The nymphs of many mayfly species share the same conformation and proportions, and a Pheasant Tail makes a serviceable imitation of them all, including the Mahogany Dun. In the weeks leading up to the hatch, fish the fly deep; when duns are on the water, fish it shallow, as a dropper behind a dry fly or indicator. The instructions for dressing this fly appear on pp. 21–23.

# Mahogany Dun Emerger Pattern

**PARASOL MARABOU**

| | |
|---|---|
| **Hook:** | #12-16 standard nymph |
| **Thread:** | Brown 6/0 or 8/0 |
| **Tail:** | Pheasant tail-feather barbs |
| **Rib:** | Fine copper wire |
| **Abdomen:** | Brown marabou |
| **Parasol puff:** | Medium or light gray poly yarn |
| **Parasol post:** | 8-inch length of 4X tippet |
| **Thorax:** | Brown dubbing |
| **Legs:** | Pheasant tail-feather barbs |

The poly yarn "parasol" on this pattern suspends the nymph body just below the surface and imitates the wings of a dun just starting to emerge. For the angler, it functions as a fairly visible built-in strike indicator that makes it easy to fish a subsurface emerger pattern without the use of a separate strike indicator or indicator dry fly—an advantage at times when the fish are wary. This is an effective pattern for selective trout feeding on the nymphs close to the surface. If desired, you can increase the height of the parasol post to make the body ride deeper beneath the film.

**1.** To form the parasol, clip two 1-inch lengths of poly yarn, shown at the left, and combine them into a single bundle as shown in the middle. Use the tippet material to tie a clinch knot around the center of the bundle, as shown at the right. Clip away the tag end of the clinch knot.

**2.** Fold the two halves of the yarn together, and hold them at the base. Run a dubbing needle or comb through the fibers several times to separate them and fluff them out.

**3.** Mount the thread behind the hook eye, and wrap a foundation to the midpoint of the shank. Position the thread ⅓ of a shank length behind the eye. Mount the parasol atop the shank. The height of the finished monofilament post—that is, the distance between the mounting wraps and yarn puff—can be varied; here, it is about ⅛ inch. Clip and bind the excess mono.

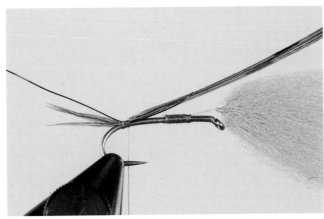

**4.** Wrap the thread to the tailing point. Align the tips of 3 to 5 pheasant tail fibers, and mount them to make a tail one hook gap in length. Do not clip the excess. Mount a length of ribbing wire over the rearmost tailing wraps. Do not clip the excess.

**5.** Align the tips of 5 to 8 marabou barbs; clip ¼ inch off the tip end. Mount them atop the rearmost tail-mounting wrap, and bind the excess marabou, along with the excess tail and rib material, to the midpoint of the shank. Clip and bind the material tags.

**6.** Gently draw all the marabou barbs into a bundle, and twist them clockwise (when viewed from above) to form a chenille-like strand. Begin wrapping the abdomen.

**7.** Continue wrapping the abdomen to the midpoint of the shank, pausing as needed to twist the strand and keep the barbs consolidated. Tie off and the clip the excess. Spiral the ribbing forward, in the opposite direction from the one in which you wrapped the marabou, taking 5 to 6 turns over the abdomen; tie it off directly in front of the abdomen, and clip the excess.

**8.** Dub a thorax to the rear of the monofilament post. Raise the post perpendicular to the shank, and take 2 or 3 wraps of dubbing at the front base of the mono to hold it in a vertical position. Dub the rest of the thorax to within 6 thread wraps' distance of the hook eye.

**9.** Mount 3 pheasant tail barbs along each side of the hook shank to form legs about one hook gap long, as shown here from underneath.

**10.** Clip and bind the excess leg material. Finish the head of the fly. Draw the poly yarn upward, and trim it to about ⅛ inch in height.

# Mahogany Dun Pattern

## MAHOGANY THORAX DUN

*Originator: Mike Lawson*

**Hook:** #12-16 standard dry-fly
**Thread:** Brown 8/0
**Tail:** Dark dun dry-fly hackle barbs or Microfibetts
**Body:** Light reddish brown dubbing
**Wing:** Dun turkey flat
**Thorax hackle:** Medium to dark dun dry-fly

This trim pattern lands lightly on the surface, rides low, and presents a distinct, well-defined wing silhouette to the trout—all of which make it ideal for fishing the slow, flat water where the duns usually emerge. But it still floats well enough to fish rises you may see in choppier water or to use as a point fly supporting a nymph dropper. Despite its somewhat delicate appearance, the fly is very durable. In choosing turkey flats for the wing, select feathers with squared-off or symmetrically angled tips and unbroken barbs.

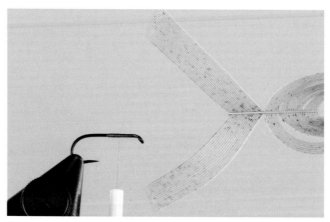

**1.** Mount the thread behind the hook eye, and wrap a foundation to the midpoint of the shank. Position the thread ⅓ of a shank length behind the eye. Select a turkey feather, and clip the feather stem 1½ to 2 shank lengths below the tip, leaving a V in the tip as shown. Preen the lower barbs toward the base of the feather, leaving a section of barbs on each side of the stem near tip; each of these sections should be one hook gap in width.

**2.** Carefully fold the 2 sections of barbs together so that the tips are aligned and the barb sections lie flat against one another to form a "panel"; they should not be crushed into a bundle.

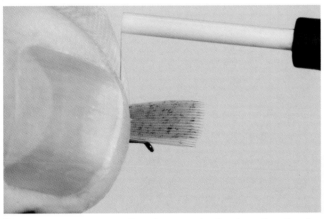

**3.** Position the wings edgewise atop the hook so that the tips extend forward beyond the tying thread a distance of one shank length. Spin the bobbin counterclockwise (when viewed from above) to flatten the thread. Raise the thread vertically on the near side of the shank, and slip it between your thumb and the panel of barbs.

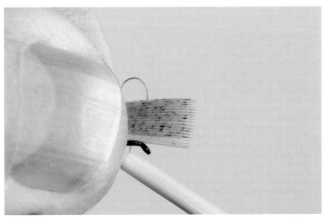

**4.** While maintaining your pinch on the wing, form a small loop over the upper edge of the barbs. Keep the loop open as you slip the thread downward between your index finger and the far side of the wing.

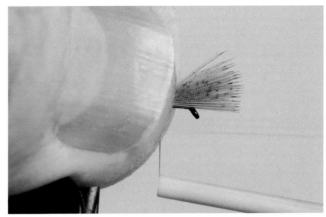

**5.** Pinch the wing firmly between your left thumb and forefinger. Pull straight down on the bobbin to tighten the loop and cinch the wing to the shank.

**6.** Repeat steps 3–5 two more times, placing each new wrap directly atop the previous one. Release the wing to check its position. The wing should sit edgewise atop the shank and maintain its flat, panel-like appearance.

**7.** When the wing is suitably positioned, bind down the butts, and clip the excess. Raise the wing vertically, and build a bump of thread abutting the front base of the wing to stand it upright. Wrap the thread to the tail-mounting point.

**8.** Using 3 to 4 Microfibetts per side, form split tails one shank length long as described for the *Baetis* CDC Compara-dun, steps 3–7, pp. 11–12. Dub a slender, slightly tapered body to the midpoint of the shank.

**9.** Prepare and mount a hackle as described for the Little Green Drake, steps 4–6, p. 42. Position the thread 5 thread wraps' distance behind the hook eye.

**10.** Wrap the hackle to the hanging thread, and tie off the feather as shown in the instructions for the Little Green Drake, steps 7–10, pp. 42–43.

**11.** Clip the excess feather tip, and finish the head of the fly. Clip away the hackle barbs beneath the hook shank, as shown in this front view.

# Mahogany Dun Spinner Pattern

**GATHERED-HACKLE MAHOGANY SPINNER**

| | |
|---|---|
| **Hook:** | #12-16 standard dry-fly |
| **Thread:** | Dark brown 8/0 |
| **Tail:** | Dun Microfibetts |
| **Body:** | Fine reddish brown dubbing |
| **Hackle:** | Light dun dry-fly |

Like the Mahogany Thorax Dun, this spinner imitation lands gently and floats lightly on the surface, which are big advantages in the delicate presentations necessary at times on the glassy water where these spinners sometimes fall. As noted earlier, Mahogany Dun spinners may keep their wings upright for a while after falling to the surface, and the arc of hackle barbs above the shank credibly imitates those adults. The hackle also makes for better-than-average visibility on a spinner pattern. Except for the colors of the materials, this fly is identical to the *Callibaetis* Gathered-Hackle Spinner and is tied using the instructions on pp. 92–93.

Here's a front view of the finished fly.

# Pale Evening Dun

What anglers refer to as Pale Evening Duns actually include twelve Western species from three genera with enough in common to be grouped together. The nymphs of these mayflies inhabit streams with reasonably clean, rocky bottoms and can be found in currents of virtually all speeds, though they tend to favor water with a moderate flow. Populations of these insects vary widely from stream to stream, from one section to another in the same stream, and from year to year. Daily hatches are generally short in duration, but the emergence periods generally last a couple of weeks. During the summer months, the emergence cycles of the various species may follow one another in sequence, or even overlap, giving the impression of a continuous, extended hatch. Generally, there aren't a large number of adults on the water at a given time. But because this group of mayflies is widely distributed through the West and the hatch period is so long, they are important to trout, and thus to anglers, on many rivers. The patterns used to imitate these mayflies are very similar to those used for Pale Morning Duns, and many anglers use the same patterns for both hatches.

| | |
|---|---|
| **Other common names:** | Gray Fox, Western Light Cahill |
| **Family:** | Heptageniidae |
| **Genera:** | *Heptagenia*, *Leucrocuta*, and *Nixe* |
| **Emergence:** | June to September, midafternoon during cooler weather and in the evening on warmer days |
| **Spinnerfall:** | Late afternoon during cool weather and in the evening during warm weather |
| **Body length:** | ¼–⅝ inch (5–15 mm) |
| **Hook sizes:** | #12-16 |

**Above:** *Pale Evening Dun nymph: brown to olive-brown.*

**Top:** *Female Pale Evening Dun: yellow-brown to cream with darker banding on top, lighter shades of top color underneath; pale gray or cream wings, occasionally with yellow tint.*

**Bottom:** *Female Pale Evening Dun (underside): lighter shade of top color.*

**Top:** *Female Pale Evening Dun spinner: shades of yellow to tan to reddish brown; clear wings, sometimes with yellow highlights.*

**Bottom:** *Female Pale Evening Dun spinner (underside): lighter shade of top color.*

# Important Fishing Stages

**Nymph** Flattened bodies, stout legs, and strong claws allow Pale Evening Dun nymphs to move among the rocks in flowing waters with little chance of being swept away. They are not commonly found drifting in the current, except for a few days prior to a hatch—when the mature nymphs migrate out of faster flows into the slower edge waters of pools, runs, and eddies—and again at emergence, when the nymphs swim to the surface. A nymph pattern is most productive during these times when the naturals are moving around. Before the hatch, fish a weighted nymph pattern close to the bottom below riffles, in runs, and especially in current seams or deep waters close to bankside shallows. During an emergence, fish a lightly weighted nymph pattern below an indicator or dry fly in the slow, shallow edge waters. From midsummer to mid-fall, a Pale Evening Dun nymph pattern is very effective tied on as a dropper below a grasshopper imitation and fished along the stream edges. It's surprising how many trout will pass up the hopper and take the nymph, especially in heavily fished waters.

**Emerger/dun** At emergence, the nymphs swim upward, and the duns shed their nymphal skins a few inches below or on the surface. The length of time the duns spend on the surface varies. They fly off quickly in warmer weather, and on such days, the trout tend to focus more on the nymphs and emergers. The duns linger on the surface during cooler weather, and the trout feed equally on nymphs, emergers, and duns. One of the best ways to fish the hatch in any weather is to use a nymph or emerger pattern as a dropper behind a dun imitation. Generally speaking, the calm waters from which these mayflies emerge lie within a rod length or two of the bank, in water depths from a few inches to a few feet, or in the slow, shallow tailouts of a pool. Trout can be extremely wary in these waters. Take time to observe the stream, especially in the evening, to avoid spooking fish.

**Spinner** Mating swarms collect along the banks of the stream. After mating, the females spend a short time in streamside foliage before flying out to nearby riffles, dipping to the surface a few times to deposit their eggs, and falling spent to the water. Unless the hatch is very heavy, the adults do not fall to the water en masse, though this does happen occasionally. Ordinarily, the spinnerfalls are relatively sparse, and the trout wait for the spent flies in slower waters—tailouts, current seams, and eddies—where the spinners are carried by the current and accumulate. You will frequently see emerging duns and spent spinners on the water at the same time, particularly in tailouts and eddies. Under these circumstances, a dun pattern generally takes fish, but if the dun is rejected or you see noses taking invisible insects, a spinner pattern is your logical recourse.

# Pale Evening Dun Nymph Pattern

**THIN SKIN MAYFLY NYMPH**

| | |
|---|---|
| **Hook:** | #12-16 standard nymph |
| **Weight (optional):** | Lead or nontoxic wire |
| **Thread:** | Brown 6/0 or 8/0 |
| **Tail:** | Mottled brown hen-hackle barbs |
| **Abdomen:** | Mottled Oak Natural Thin Skin |
| **Wing case:** | Mottled Oak Natural Thin Skin |
| **Thorax:** | Dark brown dubbing |
| **Legs:** | Brown Buggy Nymph Legs |

The weighted version of this nymph pattern sinks quickly for fishing in deeper waters, but even unweighted, the fly easily penetrates the surface film and sinks quickly enough to be used as a dropper below a dry fly. The rubber legs on the pattern do not offer much in the way of mobility or movement to the fly, but they do strongly suggest the thick, conspicuous legs of the natural. To better imitate the shape of these flat-bodied nymphs, form the underbody with thread wraps taken under very tight tension. Then flatten the thread underbody with a pair of needle-nose pliers, and saturate the thread wraps with head cement or a cyanoacrylate glue such as Super Glue or Zap-A-Gap. Let the cement dry before wrapping the Thin Skin.

**1.** If tying the weighted version of this fly, as we're doing here, weight the hook as shown in the instructions for the Pheasant Tail Nymph, steps 1–2, p. 21. Position the thread at the tailing point. Align the tips of 6 to 8 hen-hackle barbs, and mount them as a bundle atop the shank to form a tail about one hook gap in length. Use the thread to form a tapered abdomen underbody; the front end of the underbody should be just slightly smaller in diameter than the wrapped-wire thorax. Position the thread at the tailing point. Cut a 3-inch strip of Thin Skin that is ½ the hook gap in width, and taper one end to a point.

**2.** Remove the paper backing from the Thin Skin. Mount the tapered tip of the Thin Skin over the rearmost tailing wraps. Gently stretch the Thin Skin, and wrap it forward in overlapping wraps to give the abdomen a segmented appearance.

**3.** Wrap the Thin Skin up to, and then slightly beyond, the rear of the thorax. Tie it off atop the hook shank.

**4.** Fold the Thin Skin rearward, centered along the top of the shank. Wrap the thread rearward, over the fold of Thin Skin, to the front of the abdomen.

**5.** Dub the thorax about ¼ the distance to the hook eye. Mount rubber legs on each side of the shank, ahead of the dubbing, as described for the Green Drake Rubber Legs, steps 6–9, pp. 58–59, and shown in this top view.

**6.** Dub over the leg-mounting wraps and then ahead of the legs, stopping about 4 to 5 thread wraps' distance behind the hook eye.

**7.** Draw the Thin Skin smoothly over the top of the thorax, and secure it. Then clip the excess, and finish the head of the fly. Trim the legs to about one hook gap in length, or slightly longer.

# Pale Evening Dun Emerger Pattern

**ROSE'S SULPHUR EMERGER**

*Originator: Gary Rose*

| | |
|---|---|
| **Hook:** | #12-16 light wire scud |
| **Thread:** | Yellow 8/0 |
| **Tail:** | Medium-gray ostrich herl |
| **Rib:** | Yellow Krystal Flash |
| **Abdomen:** | Sulphur-yellow rabbit dubbing |
| **Wing:** | Medium-dun CDC puff feather |
| **Thorax:** | Sulphur-yellow rabbit dubbing |

CDC puffs are a wonderful material. They are simple to work with, tie in with little bulk, and give excellent flotation to a fly, though they should be treated with a powder-type floatant only. You can, however, leave the fly untreated, squeeze some water into the body and wing, and fish it sub-surface; it makes an unusually good sunken emerger imitation because the wing, even when wet, inhibits sinking and holds the fly just beneath the film. This generic, workmanlike design is highly adaptable; by changing the hook size and the colors of components, it can be used to imitate emergers of many mayfly species.

**1.** Mount the thread behind the hook eye, and lay a thread foundation on the shank to a point about halfway around the bend. You may find it easier to tilt the hook forward in the vise jaws to put the bend of the hook upward. Mount one ostrich herl to form a tail about one hook gap in length. Mount one strand of Krystal Flash atop the rearmost tailing wrap. Clip and bind the excess of both materials. Return the thread to the tail-mounting point

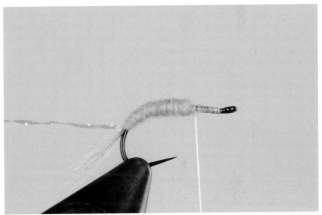

**2.** Dub a slightly tapered abdomen over the rear ⅗ of the hook shank. When the abdomen is about half wrapped, return the hook to the normal position in the jaws. Position the thread directly ahead of the abdomen.

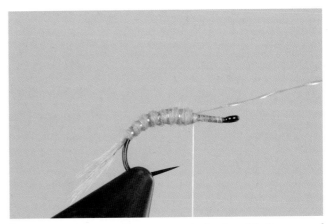

**3.** Rib the abdomen in an open spiral using 6 to 8 wraps of Krystal Flash. Tie off the rib.

**4.** Clip the excess rib material, and position the thread at the front of the abdomen. Position the CDC puff atop the shank in front of the abdomen so that the barbs tips extend rearward almost to the end of the abdomen.

**5.** Secure the feather tightly with wraps that abut the front of the abdomen.

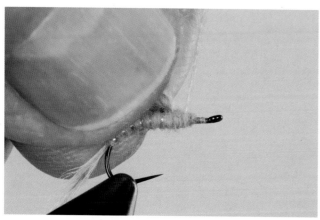

**6.** Clip and bind down the excess CDC. Preen the wing fibers rearward to keep them out of the tying field, and use a thinly dubbed thread to form a tapered thorax. Then finish the head of the fly.

# Pale Evening Dun Pattern

## PARACHUTE LIGHT CAHILL

*Originator: Version of Dan Cahill pattern*

| | |
|---|---|
| **Hook:** | #12-16 standard dry-fly |
| **Thread:** | Pale-yellow 8/0 |
| **Wing post:** | Wood-duck flank-feather barbs |
| **Hackle:** | Light ginger dry-fly |
| **Tail:** | Light ginger Microfibetts or dry-fly hackle barbs |
| **Body:** | Cream or pale-yellow dubbing |

The Light Cahill was designed to imitate an Eastern mayfly species reasonably similar in appearance to the Pale Evening Duns. This flush-floating parachute version has a lower, less aggressive silhouette than the original collar-hackled fly and is better suited to the flat, often shallow water where trout frequently feed on Pale Evening Duns. The fly floats well enough to support an unweighted nymph or emerger as a dropper. In a pinch, this pattern can also be used to imitate the Pale Evening Dun spinner; if trout refuse the fly, try trimming down the wing to within 1/8 inch of the topmost hackle wrap. This fly is dressed in a fashion similar to that shown for the Speckle-Wing Parachute, pp. 90–91, so the following instructions are abbreviated. Some tiers dress this fly with a tail formed from a straight bundle of hackle fibers; we believe this may exaggerate the body length in the eyes of the trout, making the fly appear too large. We prefer the fan-style tail used here.

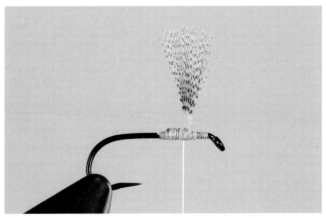

**1.** Prepare the wood-duck barbs, mount, and post up the wings as shown in the instructions for the Speckle-Wing Parachute, steps 1–2, pp. 90–91.

**2.** Prepare and mount a parachute hackle as shown in the instructions for the Parachute BWO, steps 3–4, p. 16.

**3.** Wrap the thread to the tailing point. Align the tips of 8 to 10 Microfibetts, and mount them as a bundle to form a tail one shank length long. Bind and clip the excess. To flare the tail fibers, push your thumbnail or a dubbing needle against the base of the tail from underneath. Once the fibers are flared out, you can adjust them into a symmetrical fan shape by positioning them with your fingers.

**4.** Use a sparsely dubbed thread to form a slender, tapered body, stopping 4 to 5 thread wraps' distance behind the eye.

**5.** Wrap and tie off the hackle as described for the Parachute BWO, steps 7–10, pp. 16–17. Finish the head of the fly.

# Pale Evening Dun Spinner Pattern

**GATHERED-HACKLE PED SPINNER**

| | |
|---|---|
| **Hook:** | #12-16 standard dry-fly |
| **Thread:** | Tan 8/0 |
| **Tail:** | Light dun Microfibetts |
| **Body:** | Tan dubbing |
| **Hackle:** | Light dun dry-fly |

During the warm summer months, Pale Evening Dun (PED) spinnerfalls are most commonly encountered, unsurprisingly, in the evening. The fanlike spread of hackle barbs over the shank gives this pattern some degree of visibility in low-light conditions. The clean body silhouette and overall spare design of this pattern make it particularly effective on slow, smooth currents, where the trout get a very good look at it. Except for the colors of the materials, this fly is tied exactly like the *Callibaetis* Gathered-Hackle Spinner shown on pp. 92–93.

Here's a front view of the finished fly.

# Chapter Two

# Caddisflies

## MAJOR HATCHES

## Green Sedge

Unlike a great many species of caddis, the larvae of the Green Sedge do not build cases of small stones, sand, or plant debris—an adaptation that enables their predatory habits. Unencumbered by heavy and clumsy cases, they move freely about the streambed and feed on smaller organisms. These caddis larvae are found in the riffles of cold, oxygen-rich waters, and freestone rivers with an abundance of riffles and fast, shallow, rocky runs frequently have large populations of these insects. This family has a number of species, which hatch at different times of the year, usually peaking before and after the summer months. Even though the adults emerge only sporadically through most of the hatch period, the larvae, pupae, and adults are available to the trout for an extended period of time each year, making the Green Sedge hatch an important one on many Western waters.

**Top:** Rhyacophila *nymph/larva: shades of green to tan.*

**Bottom:** Rhyacophila *nymph/larva (underside): lighter shade of top color.*

| | |
|---|---|
| **Other common name:** | Green Rock Worm |
| **Family:** | Rhyacophilidae |
| **Genus:** | *Rhyacophila* |
| **Emergence:** | May to October, with peak periods in early June and mid-September, any time from midday to late afternoon, sporadic rather than en masse |
| **Egg-laying flight:** | Late afternoon |
| **Larva body length:** | ½–¾ inch (12–18 mm) |
| **Larva hook sizes:** | #12-16, 3XL |
| **Adult body length:** | ⁵⁄₁₆–⁵⁄₈ inch (8–16 mm) |
| **Adult hook sizes:** | #10-16, with #14 most common |

119

**Left:** Rhyacophila *pupa: green to brown body; tan-brown to gray wing pads.*

**Left:** Rhyacophila *adult: mottled brown to gray wings.*

**Above:** Rhyacophila *adult (underside): green to brown body.*

# Important Fishing Stages

**Larva** These free-living caddis larvae, often referred to as Green Rock Worms, are predators. A pair of clawlike hooks at the rear of the abdomen allows these insects to grip stones and debris on the bottom as they crawl about and hunt small prey. But because they pursue their food, these larvae are exposed to the swift currents of their preferred habitat, and many are swept from the rocks. Being poor swimmers, they often drift long distances as they settle back to the streambed, and patterns imitating these larvae should be fished close to the bottom. When the larvae are mature, they enclose themselves in cocoons and pupate for about a month, during which time they are not available to the trout. So the best time to fish a larva pattern is 4 to 8 weeks before the hatch, when the larvae are almost fully grown and at their most active.

**Pupa/emerger** At the time of emergence, Green Sedge pupae leave their cocoons sporadically throughout the afternoon hours, drifting downstream as they swim to the surface. While they hang suspended just below the surface film, their pupal shucks spilt open, and the adults quickly emerge. Pupa patterns can be fished at any depth during the hatch, but unless large numbers of flies are emerging, the trout tend to stay in sheltered waters behind rocks or downstream from riffles and intercept the pupae before they reach the surface. Often the only indications that trout are feeding on these pupae are a few adults flying over the water and an occasional boil from a trout taking a pupa near the surface. The best time to fish an emerger pattern just under the surface is during the peak of emergence, when larger numbers of cripples and delayed emergers are available.

**Adult** Even though the adults typically emerge only sporadically, because of their numbers and the length of the hatch, trout are accustomed to seeing and feeding on them. And, in fact, an adult Green Sedge imitation makes an excellent searching pattern during the afternoon, when there are few or no visible rises. A highly effective way to fish this hatch is to trail a sunken pupa pattern behind the dry fly wherever the fishing regulations allow the use of two flies.

**Egg-laying female** In the late afternoon, females gather over the riffles. They land on the water and swim to the bottom or alight on an exposed rock and crawl beneath the surface; once underwater, they attach their eggs to submerged objects. After depositing their eggs, the spent females are swept downstream; some make their way to the surface again, but few leave it. Large numbers of these spent females in the water often trigger a feeding frenzy—a signal that it's time to fish a sunken adult pattern in riffles and the waters below them.

# Green Sedge Larva Patterns

## KRYSTAL FLASH GREEN ROCK WORM

*Originator: Rick Hafele*

**Hook:** #12-16 1XL–2XL nymph
**Weight (optional):** Lead or nontoxic wire
**Thread:** Brown 8/0
**Body:** 2 to 4 strands of peacock or green Krystal Flash
**Head:** Dark brown dubbing

This fly is as simple as they come and proof that a pattern need not be elaborate to be effective. To fish it deep, especially in fast pocketwater, weight the entire hook shank with wire that is about $\frac{1}{2}$ the hook diameter; thicker wire will distort the body proportions. Fish the fly dead-drift just off the bottom. The unweighted version, shown here, can be fished as a dropper behind a dry fly for shallow fishing, or behind a larger weighted nymph or split shot for deeper water. Coating the body with a thick head cement is optional but recommended to give the fly more durability. The sharp teeth of trout can easily sever the Krystal Flash, and the body will unravel.

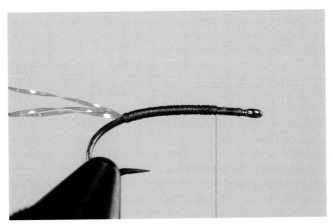

**1.** Mount the thread behind the hook eye, and wrap a smooth underbody to the rear of the shank. Mount 2 to 4 strands of Krystal Flash as a bundle. Bind and clip the excess. Position the thread $\frac{1}{5}$ of a shank length behind the eye.

**2.** Gather the Krystal Flash strands into a bundle. Twist them clockwise (when viewed from above) to form a single strand, and wrap it forward to create a slender body over the rear $\frac{4}{5}$ of the shank.

**3.** Tie off the Krystal Flash, and clip the excess; coat the body with a thick cement. Use a thinly dubbed thread to form a head slightly larger in diameter than the body, and finish the fly. Use a dubbing needle to pick out a few fibers on the sides of the head to suggest legs, as shown in this top view; trim any excessively long fibers.

## STRETCH LACE ROCK WORM

| Hook: | #12-14 TMC 200R |
|---|---|
| Weight: | .010-inch lead or nontoxic wire |
| Thread: | Chartreuse 6/0 or 8/0 and dark brown 8/0 |
| Underbody: | Chartreuse tying thread, colored with green Sharpie permanent marker |
| Abdomen: | Small Chartreuse Ice Stretch Lace |
| Shellback: | Brown ⅛-inch Scud Back |
| Thorax: | Green Antron dubbing |

With its distinctly segmented, two-toned abdomen and glossy shellback thorax, this fly is a more realistic imitation of the natural than the Krystal Flash Green Rockworm. The translucence of the Ice Stretch Lace allows the underbody colors to show through; the permanent marker used to tint the thread is protected by the body material, so it won't fade with use as a tinted exterior surface would. Fish the weighted version of the pattern (shown in the following sequence) in deeper or faster waters. You can also tie an unweighted version on a size 16 hook, and use it as a dropper when fishing shallow edge waters. When tying the unweighted version, mount the Ice Stretch Lace ¼ of a shank length behind the eye; wrap rearward, forming a smooth thread underbody as you bind the Stretch Lace atop the shank. Then wrap forward to form the body.

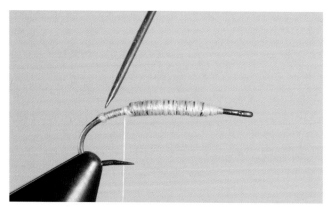

**1.** Wrap the wire over the middle ⅔ of the hook shank, and secure it with the chartreuse thread as described for the Pheasant Tail Nymph, steps 1–2, p. 21. Wrap the thread to the rear of the hook shank, and make a small thread bump. Then wrap the thread forward to the rear of the wire wraps. Coat the thread bump with head cement.

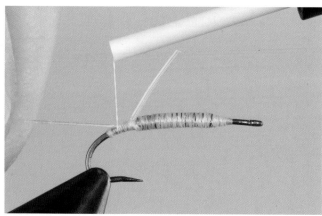

**2.** Use 4 tight thread wraps to mount the strand of body material on top of the hook shank; leave the tag end long enough to grip. While keeping the body material on top of the hook shank, wrap the thread rearward, gradually increasing the tension on the strand to stretch it and decrease the diameter. Stop the thread about 2 thread wraps' distance in front of the thread bump.

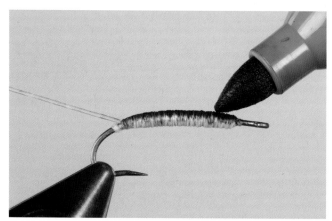

**3.** Stretch the tag end of the body strand, and trim it at the base. Flatten the thread by spinning the bobbin (counterclockwise when viewed from above), and form a smooth, tapered underbody to the front of the wire wraps. Tie off and trim the thread. If you wish, at this point you can coat the thread underbody with head cement and let it dry. The cement will prevent the green marker from bleeding down the sides of the body and give a more distinct color separation. Color the top of the underbody with a green marker.

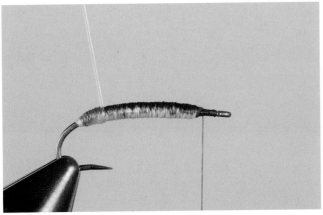

**4.** Mount the dark brown tying thread behind the hook eye, and position it in front of the wire wraps. Stretch the body strand, and take the first wrap directly in front of the thread bump. Wrap forward, decreasing the tension on the strand so that the rear of the body is tapered.

**5.** Continue wrapping the lace to the hanging thread, and secure it with 6 tight thread wraps. Wrap rearward over the top of the body so that the thread is positioned ⅕ of a shank length behind the hook eye.

**6.** Trim the excess body material. Mount a length of Scud Back on top of the hook shank with 6 tight thread wraps. Trim the excess, and bind down the butts of both the Scud Back and the body material. Position the thread at the front of the body.

**7.** Dub the thorax, stopping 4 to 5 thread wraps' distance behind the eye. Position the thread in front of the thorax.

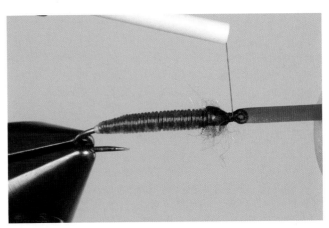

**8.** Pull the Scud Back over the top of the thorax, stretching it as necessary to make the shellback as wide as the body. Secure it with 6 to 8 tight thread wraps, as shown in this top view.

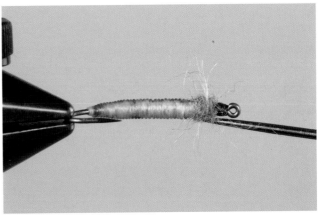

**9.** Finish the head of the fly. Pick out a few fibers from each side of the thorax to imitate the legs, as shown in this bottom view. Then trim the picked-out fibers to about a hook gap in length.

# Green Sedge Pupa Patterns

**LATEX CADDIS PUPA**

**Hook:** #12-16 2XL nymph
**Weight (optional):** Lead or nontoxic wire
**Thread:** Brown 8/0
**Abdomen:** Green or olive latex, ⅛-inch strip
**Body cement (optional):** Softex Fly Tying Coating
**Wing pads:** Dark brown duck wing-quill sections, coated with Dave's Flexament
**Thorax:** Brown Antron dubbing

Latex strips make a nicely segmented abdomen with a realistic appearance. If you can't locate colored latex, use natural latex, which is cream-colored, and tint it with waterproof marker after completing the body in step 5. Though latex is strong in some respects, when wrapped under tension, it can be cut or split by a trout's teeth; moreover, it can dry out and degrade over time, like an old rubber band, and become weak. To improve its durability and longevity, coat the latex with a thin layer of Softex after the abdomen is completed in step 5. The Softex will also help lock in any color applied with a waterproof marker. The abdomen can also be formed with another type of flat, elastic strip, such as Scud Back. Duck wing quills make the best wing-pad material; if you can't find dark brown feathers, use natural feathers and color them with a waterproof marker.

**1.** Weight the hook over the middle ⅓ of the shank, using the instructions for the Pheasant Tail Nymph, steps 1–2, p. 21. Position the thread at the rear of the hook. Taper one end of the latex strip to a point, and mount the point of the strip very tightly on the near side of the hook shank.

**2.** Use the tying thread to form a smooth underbody that tapers at both ends, and position the thread at the front of the underbody.

**3.** Take 2 wraps of the strip forward under tight tension. The second wrap should partially overlap the first.

**4.** Relax the pressure on the strip so that you are applying only moderate tension, and continue wrapping forward to about the midpoint of the shank, with each wrap partially overlapping the previous one. Relaxing the tension will increase the diameter of the abdomen and better define the body segments.

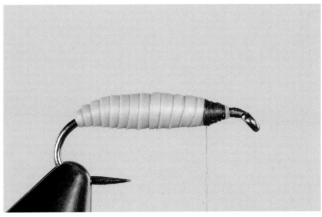

**5.** Increase the tension on the latex, and wrap forward to the tying thread. Secure the latex strip, and clip the excess. If you are using natural latex, color the abdomen at this point. Otherwise, apply a thin coat of Softex, and let the fly dry before proceeding to the next step.

**6.** From a pair of matched wing feathers, cut a pair of barb segments about ¹⁄₂ the hook gap in width. Coat them with Flexament to prevent splitting, and let them dry.

**7.** Position one wing-quill section against the near side of the shank to form a wing pad about ¹⁄₂ the length of the abdomen. The quill section should curve inward, with the tip sweeping downward.

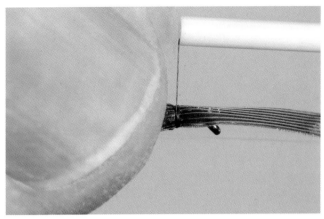

**8.** Maintaining the position of the quill section, pinch it in your left fingers, and mount it on the near side of the shank.

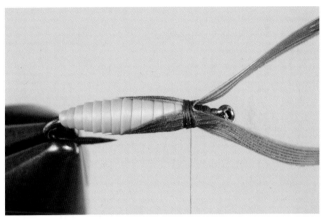

**9.** Return the thread to the rearmost mounting wrap, and mount the second section of quill on the far side of the shank so that it matches the first section in length, as shown here from the top.

**10.** Clip and bind the excess, and return the thread to the rearmost mounting wrap. Dub the thorax, and finish the fly. Use a dubbing needle or teaser to pick some fibers from the underside of the head. Trim any excessively long fibers.

## DEEP SPARKLE PUPA

*Originator: Gary LaFontaine*

| | |
|---|---|
| **Hook:** | #12-16 1XL nymph |
| **Weight:** | Lead or nontoxic wire |
| **Thread:** | Brown 8/0 |
| **Overbody:** | Tan or olive Antron yarn |
| **Underbody:** | Green Antron dubbing |
| **Legs:** | Brown hen or partridge feather barbs |
| **Head:** | Brown dubbing |

The capsule-like Antron overbody imitates the pupal shuck and gives this fly some shimmer underwater, but keep this overbody sparse so that the underbody shows through. This pattern can be fished anywhere in the water column, but it is particularly effective when presented deep, just off the bottom. Fish the fly dead-drift, and at the end of the drift, let the fly swing on a tight line until it is directly below you; it will rise up off the bottom like a natural pupa ascending to emerge. By changing the underbody color, you can use this same design to match other species of caddis pupae. This is our top choice for a caddis pupa pattern.

**1.** Cut two 2-inch lengths of Antron yarn. Depending on the hook size, you may need to remove some fibers from each strand to avoid producing too dense an overbody. Comb through half of the yarn strand to separate the fibers; then reverse the strand, and comb through the other half. Repeat with the second strand, and set them both aside.

**2.** Weight the middle half of the hook shank as described for the Pheasant Tail Nymph, steps 1–2, p. 21. Position the thread at the rear of the shank. Mount one strand of yarn on the far side of the hook shank with 4 tight thread wraps.

**3.** Mount the second strand of yarn on the near side of the shank, as shown in this top view. Clip the excess yarn, and bind down the butts.

**4.** Dub the body over the rear ¾ of the hook shank. Position the thread at the front of the body.

**5.** Pull the yarn strand on the near side forward along the side of the hook shank. Try to keep the fibers spread rather than bunched together. When the strand is in position, secure it with 2 firm thread wraps. Do not clip the excess.

**6.** Repeat step 5 with the strand on the far side, again securing it in front of the body with 2 firm wraps, as shown in this top view. Do not clip the excess.

**7.** Use a dubbing needle to lift the fibers slightly and pull them sideways to fill any sparse areas. Work around the hook shank on all sides to make a bubble-like capsule surrounding the body.

**8.** Once the bubble is formed, secure the yarn strands tightly. Clip and bind the excess. At the front end of the bubble, clip a few fibers of yarn from the top and sides; smooth them rearward, like a tail, as shown.

**9.** Align the tips of 4 to 6 hen-hackle or partridge barbs. Mount them on the far side of the shank to make legs about ¾ of a shank length long. Mount a second bundle of barbs on the near side, as shown in this top view.

**10.** Bind and clip the excess leg material. Dub the head, and finish the fly.

# Green Sedge Emerger Patterns

## PARTRIDGE AND GREEN

| | |
|---|---|
| **Hook:** | #12-16 standard nymph |
| **Thread:** | Olive 8/0 |
| **Abdomen:** | Green floss |
| **Thorax:** | Tan rabbit dubbing |
| **Hackle:** | Brown partridge |

This classic fly design is custom-made for fishing the Green Sedge hatch when trout are keying on pupae swimming to the surface to hatch. If you see flashing trout or subsurface boils, cast the fly quartering downstream, and let it swing across the current directly in front of the fish. This same down-and-across swing is an excellent way to search water during the hatch season when you see no visible feeding activity. Resist the temptation to overhackle the fly. On the swing, water resistance forces the hackle barbs back around the body like a shuck, and you want the body color to be visible through the surrounding veil of barbs.

**1.** Prepare the hackle feather by stripping one shank length's worth of barbs from one side of the feather, as shown.

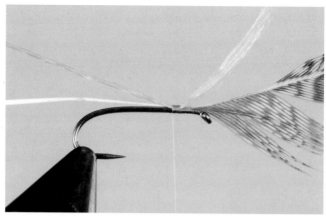

**2.** Mount the thread behind the hook eye. Position the thread 4 thread wraps' distance behind the eye. Position the feather atop the shank with the stripped portion of the stem facing upward. Bind the feather stem, just below the lowermost barbs, to the top of the shank with 3 to 4 tight wraps toward the rear. Then mount a 6-inch length of floss atop the shank, as shown.

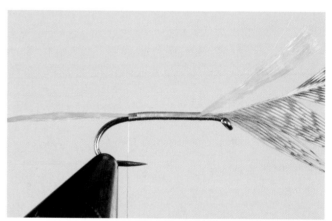

**3.** Flatten the thread by spinning the bobbin counterclockwise (as seen from above), and wrap to the rear of the hook shank, binding both the feather stem and floss to the shank as you go. Spin the bobbin again as necessary to keep the thread flat and the underbody smooth. When you are a few thread wraps' distance from the rear of the hook, trim the feather stem, and take 2 to 3 more thread wraps to the rear, binding the floss.

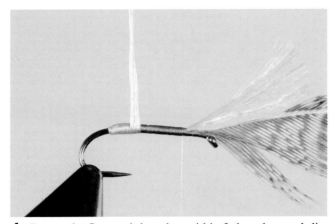

**4.** Return the flattened thread to within 3 thread wraps' distance of the original floss-mounting point, again keeping the underbody smooth. Draw the floss downward, and smooth it into a flat, ribbonlike band. Begin wrapping the floss forward, with each wrap slightly overlapping the previous one. Keep the floss in a flat band. If some floss filaments start to stray or separate from the band, pause and twist the floss clockwise to reconsolidate the fibers.

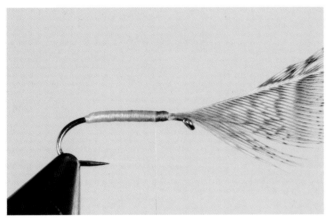

**5.** Wrap up to the tag of floss that was left from mounting. Tie off the floss strand, clip the excess, and clip the original tag of floss. Wrap the thread back over the floss to a point ⅓ of a shank length behind the eye.

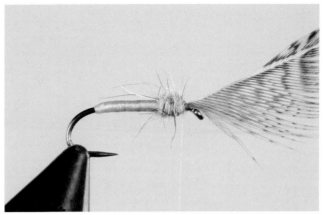

**6.** Dub a thorax about twice the thickness of the abdomen to the hackle tie-in point. Position the thread at the front of the thorax.

**7.** Clip the feather tip in a pair of hackle pliers, and take 2 turns rearward, toward the thorax.

**8.** Take 2 turns of thread over the hackle tip. Then take 2 to 3 wraps of thread forward through the wrapped hackle; as you make these wraps, wiggle the thread from side to side to prevent binding down any hackle barbs. When you reach the hook shank directly ahead of the frontmost hackle wrap, form a small thread head, and finish the fly. Then clip away the excess hackle tip.

## PARACHUTE CADDIS EMERGER

*Originator: Shane Stalcup*

| | |
|---|---|
| **Hook:** | #12-16 fine-wire scud |
| **Thread:** | Brown 6/0 |
| **Body:** | Green or olive Hare's Ice Dub |
| **Wing post:** | White poly yarn |
| **Wing:** | Dark dun Medallion sheeting |
| **Hackle:** | Dun dry-fly, one size larger than normally used for the hook size |
| **Thorax:** | Green or olive Antron dubbing |

As a rule, we don't care much for winging trout flies with synthetic film materials. Aside from being prone to split, they often produce enough wind resistance during casting to make the fly spin and twist the tippet. But these problems are minimized by the shorter wings tied in the down-wing style on this pattern, and the material does offer lifelike venation and translucence. The white wing post makes this fly easy to see, and it floats well even in the choppy waters where Green Sedges emerge.

**1.** Mount the thread behind the hook eye. Wrap a thread foundation to the midpoint of the shank, and return the thread halfway to the hook eye. Mount a poly yarn wing post as shown in the instructions for the Parachute BWO, steps 1–2, p. 15.

**2.** Wrap the thread to the rear of the hook, and dub the abdomen over the rear $^2/_3$ of the hook shank. Position the thread at the front of the wing post.

**3.** Cut a 2-inch strip of Medallion sheeting that is about $^2/_3$ the hook gap in width. Using crisscross wraps, mount the wing crosswise atop the shank directly in front of the wing post.

**4.** Fold one strip of the wing material along each side of the body. Wrap the thread rearward over the base of the wings, continuing past the wing post, securing the material until you reach the front of the abdomen.

**5.** Trim the wings at an angle so that they extend just beyond the end of the body. Prepare and mount the hackle feather as described for the Parachute BWO, steps 3–4, p. 16.

**6.** Dub the thorax of the fly to within 3 to 4 thread wraps' distance from the hook eye.

**7.** Wrap the hackle, taking 5 to 6 wraps down the post, and tie it off as described for the Parachute BWO, steps 7–10, pp. 16–17. Clip the hackle tip and finish the fly. Trim the wing post to about one hook gap in height.

# Green Sedge Adult Patterns

## DEER HAIR CADDIS

| | |
|---|---|
| **Hook:** | #12-16 standard dry-fly |
| **Thread:** | Brown 6/0 |
| **Body hackle:** | Dun dry-fly |
| **Body:** | Green or olive dubbing |
| **Wing:** | Deer hair |

The deer-hair wing on this version of the venerable Elk Hair Caddis better imitates the wing color of the Green Sedge, as it does the wing colors of a number of Western caddis species. The trimmed hackle beneath the fly puts the body closer to the surface, makes the fly more stable, and helps it land dependably upright on the water. This is still a high-riding pattern, however, that imitates a fluttering adult. It's buoyant enough for rough water or fishing with a weighted pupa or diving adult imitation as a dropper. This is a fly-box standard for the Green Sedge hatch. As always, choosing deer hair for a particular pattern is a trial-and-error process; for this fly, look for hair that flares out at about 45 degrees under thread pressure. More flare than that makes the fly too bushy; less flare indicates that the hair is nearly solid and lacks buoyancy.

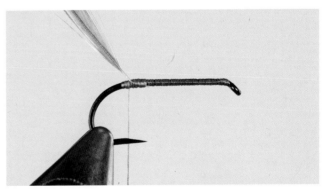

**1.** Mount the thread behind the hook eye, and wrap a thread foundation rearward, stopping about 3 thread wraps' distance ahead of the normal tailing point on the hook. Prepare and mount a hackle as described for the Little Green Drake, steps 4–6, p. 42. When mounting the feather, leave a length of bare hackle stem between the rearmost mounting wrap and the lowermost barbs on the feather. The length of this bare stem is equal to ½ the thickness of the dubbed abdomen. Position the thread just behind the hackle.

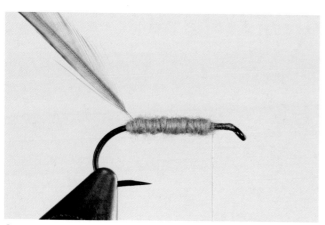

**2.** Beginning behind the mounted hackle, dub the abdomen over the rear ¾ of the shank. Position the thread directly ahead of the abdomen.

**3.** Grasp the feather tip in a pair of hackle pliers, and wrap the hackle forward in an open spiral over the abdomen. Keep the dull (back) side of the feather toward the hook eye, the feather barbs perpendicular to the hook shank, and the spacing of the wraps uniform.

**4.** Continue wrapping forward until you reach the front of the abdomen. At the tie-off point, hold the hackle feather above the shank and angled slightly toward the hook eye. Unwrap the thread until it is right next to the last wrap of hackle.

**5.** Transfer the hackle pliers to your right hand. With your left hand, take a tight wrap of tying thread over the hackle stem directly ahead of the last hackle wrap. Once the hackle is secured with this wrap, release the hackle tip, and take 2 to 3 tight wraps forward, binding down the feather.

**6.** Clip the excess feather, and position the thread directly ahead of the frontmost hackle wrap. Clean and stack a bundle of deer hair about ½ the thickness of the hook gap, as described for the Sparkle Dun, steps 1–4, pp. 27–28. Position it atop the shank so that the hair tips extend to the hook bend.

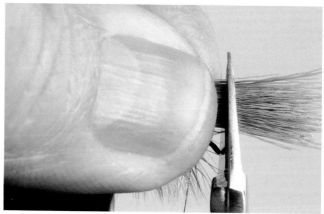

**7.** Hold the bundle in position, and transfer it to your left fingers. Trim the hair butts even with the front of the hook eye.

**8.** Secure the hair to the top of the shank with 5 to 7 tight thread wraps. As you wrap, use your left fingers to hold the hair firmly atop the shank and resist the thread torque that would otherwise draw the hair down the far side of the shank.

**9.** To further secure the hair, angle the thread forward and take a wrap through the rear section of the clipped hair butts, wiggling the thread as you wrap to slip it between the hairs. Take a second, identical wrap through the middle of the hair butts, and then a third through the front section of the butts.

**10.** Finish the head of the fly ahead of the hair butts, around the shank only. Trim the hackle barbs beneath the shank so that they are even with the hook point.

## SNOWSHOE CANOE FLY

| | |
|---|---|
| **Hook:** | #12-16 standard dry-fly |
| **Thread:** | Brown 6/0 or 8/0 |
| **Body:** | Green or olive Antron dubbing |
| **Wing:** | Medium or dark dun snowshoe hare's foot hair |

Snowshoe hare's foot hair makes this version of the Canoe Fly virtually unsinkable. Simple to tie and almost indestructible, this pattern imitates a crippled or spent adult and is well suited to fishing the broken water and pockets where Green Sedges are often found. On smoother currents, this fly sits flush on the water and creates a lifelike footprint in the surface film. Snowshoe hare's foot hair floats a fly extremely well, so you don't need a thick bundle of fibers to dress the wing. Too much material is difficult to secure to the hook, and the resulting volume in the wing produces an unnatural silhouette.

**1.** Mount the thread behind the hook eye, and wrap a thread foundation to the rear of the shank. Dub the body over the rear $4/5$ of the hook shank. Position the thread directly ahead of the body.

**2.** Clip and clean a bundle of snowshoe hare's foot hair as shown in the instructions for the BWO Snowshoe Dun, steps 1–2, p. 13. Mount the bundle atop the shank as described for the PMD Hare-Wing Dun, steps 6–7, p. 32.

**3.** Finish the fly behind the hook eye, around the shank only. Clip the front bundle of snowshoe hare, following the angle of the hook eye, to make a tuftlike head. Clip the wing hairs so that they extend just beyond the rear of the body.

# Green Sedge Egg-Laying Patterns

**DIVING CADDIS**

*Originator: Gary LaFontaine*

**Hook:**                      #12-16 1XL nymph
**Weight (optional):** Lead or nontoxic wire
**Thread:**                Dark brown 6/0 or 8/0
**Body:**                   Green or olive Hare-Tron dubbing
**Underwing:**        Grouse or mottled brown hen-feather barbs
**Overwing:**          White Antron yarn
**Hackle:**              Brown wet-fly

Female Green Sedges can be quite active underwater, swimming downward to lay their eggs and sometimes returning the surface. Many females, however, are swept away by the current and drift helplessly downstream. Rig this pattern beneath an indicator, with split shot if the fly is not weighted. By fishing the fly drag-free through part of the drift, and then swimming it back to the surface, you can imitate both behaviors of the natural insect. Keep the fly sparse, particularly the hackle, to promote quick sinking and movement in the water. You can use hen hackle or, as we prefer, soft, webby, poor-quality rooster hackle.

**1.** Mount the thread behind the hook eye, and wrap a thread foundation to the rear of the shank. Dub the body over the rear ¾ of the shank, and position the thread at the front of the body.

**2.** Align the tips of 20 to 30 grouse-feather barbs, clip them, and consolidate them into a bundle. Mount them atop the shank directly ahead of the body so that the tips extend just beyond the hook bend. Clip and bind the excess. Position the thread at the rearmost wing-mounting wrap.

**3.** Clip a sparse bundle of Antron yarn. Mount it atop the grouse barbs to form an overwing that extends just beyond the barb tips. Clip and bind the excess. Position the thread at the rearmost wrap used to mount the Antron.

**4.** Prepare and mount the hackle as shown in the instructions for the Little Green Drake, steps 4–6, p. 42, except that the hackle here should be mounted with the top, or shiny, side of the feather facing upward. Position the thread about 3 thread wraps' distance from the rearmost wrap used to mount the hackle.

**5.** Clip the tip of the feather in a pair of hackle pliers, and take one wrap around the shank directly in front of the body.

**6.** With your left fingers, draw the wrapped hackle barbs rearward, and take a second wrap of the feather directly ahead of the first.

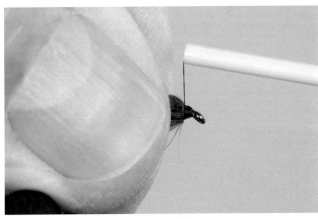

**7.** Tie off the hackle as described for the Little Green Drake, steps 9–10, pp. 42–43. Trim the feather tip. Draw all the hackle barbs rearward, and form a tapered head that slightly overlaps the base of the last hackle wrap to slant the hackle barbs toward the rear of the fly. Then finish the head of the fly.

## CDC AND WIRE CADDIS

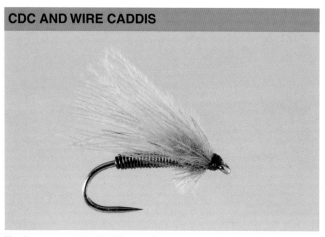

| Hook: | #12-16 1XL nymph |
|---|---|
| Thread: | Brown 6/0 or 8/0 |
| Body: | Green or olive wire, no larger than diameter of hook wire |
| Wing/legs: | 1 or 2 dun CDC feathers |

Because it provides excellent flotation, CDC is most commonly used on dry flies. But it is also a superb material for subsurface patterns: the supple barbs are quite mobile underwater, and they trap tiny air bubbles that shimmer and glisten. The weight of the wire body on this pattern easily overcomes the buoyancy of the CDC and allows the fly to penetrate the surface film easily and sink fairly quickly. In deeper water, however, you may need to add split shot to the leader. Fish the fly drag-free; at the end of the drift, let the fly swing on a tight line until it's directly below you to imitate a female returning to the surface. In choosing the body material, look for a wire that is no larger in diameter than the hook wire, and preferably a little smaller.

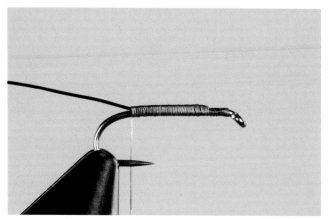

**1.** Mount the thread behind the hook eye, and wrap a foundation ¼ of the distance to the rear of the shank. Cut a length of wire for the body. Mount it atop the shank, and bind it down as you wrap back to the rear of the hook.

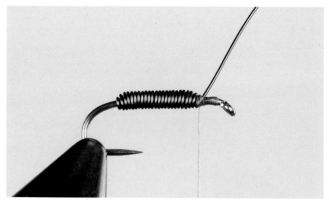

**2.** Wrap the thread forward to the end of the wire tag atop the shank. Wrap the wire forward over the rear ¾ of the shank; secure it when you reach the tying thread.

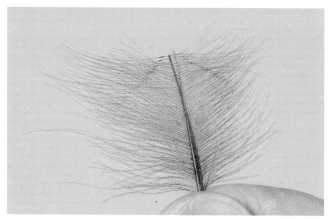

**3.** Clip the excess wire. Wrap the thread halfway to the hook eye. Prepare a CDC feather by first stripping away the short, scraggly fibers at the base. Holding the feather by the tip, preen the barbs toward the base of the feather so that they stand out perpendicular to the stem. Then trim away the tip of the feather. For small flies, one feather may suffice. For larger hooks, choose a second feather similar to the first, and prepare it the same way. Then stack the 2 feathers so that the curvatures match and the stems are aligned, as shown here. Set them aside in a handy spot.

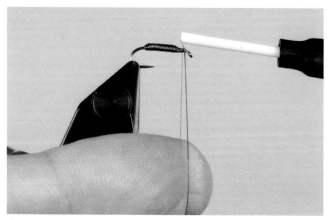

**4.** Form a dubbing loop by stripping about 8–9 inches of thread from the bobbin. Hold your left index finger about 3–4 inches beneath the hook shank. Pass the thread around your finger and back to the hook shank behind the eye, as shown.

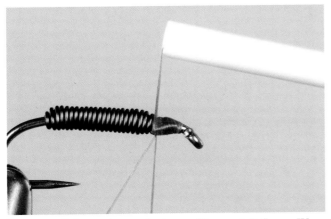

**5.** Keep slight tension on the loop with your left finger. Wrap the thread rearward, binding the 2 loop threads to the shank, until you reach the front of the wire body.

**6.** Position the thread behind the hook eye. Hang a dubbing hook, twister, or whorl from the bottom of the loop. Wax both of the loop threads with dubbing wax.

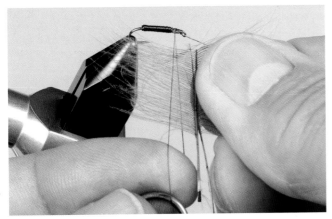

**7.** Insert the barbs on the left side of the feather between the loop threads so that the feather stem is parallel to the threads. The distance between the threads and the feather stem should be about ¼ inch.

**8.** Carefully slip your index finger out of the loop. As you do so, use your left thumb and middle finger to pinch the loop threads against the tip of the dubbing hook or twister. Pull the threads gently downward with these fingers to close the loop and trap the CDC feather between the loop threads. With scissors, trim the barbs close to the feather stem.

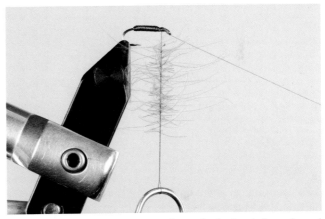

**9.** Spin the dubbing hook or twister clockwise (when viewed from above) until the CDC barbs are locked between the threads and flare outward in a chenille-like strand.

**10.** Wrap the dubbing loop around the shank, directly ahead of the abdomen. The first few wraps will be only the bare thread at the top of the loop; keep wrapping, directly in front of the abdomen, until you've taken one full turn of the flared CDC fibers around the shank.

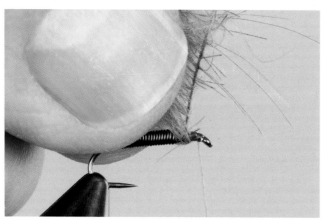

**11.** Raise the dubbing hook or twister vertically. With the tips of your left thumb and forefinger, reach around the vertical thread, and gently draw the CDC barbs upward and rearward at a 45-degree angle.

**13.** Repeat steps 11–12, drawing the CDC barbs upward and rearward with each wrap. Continue repeating these steps—forming a second dubbing loop if necessary—until you reach the hook eye.

**12.** Take another wrap of the dubbing loop, letting the wrapping tension pull the CDC barbs from your fingertips as you wrap, and ending with the dubbing loop again held vertically.

**14.** Tie off the dubbing loop behind the hook eye, clip the excess, and finish the fly. Trim the wing fibers to be even with the rear of the hook. Then trim away any CDC fibers beneath the shank, as show in this front view.

# Mother's Day Caddis and Black Caddis

In the West, the genus *Brachycentrus* includes two highly important species of caddisflies. One of them, often referred to as the Mother's Day Caddis, hatches in the spring, and the adults generally have dark brown wings. The other species emerges in mid to late summer; the adults have dark gray to black wings, and most anglers simply call them Black Caddis. The Mother's Day Caddis is the first caddis super hatch of the year and tends to be the more spectacular of the two emergences. But because it occurs just before or at the start of the spring runoff, it's wise to check the stream conditions before heading out. High, off-color water can make the hatch unfishable. The Black Caddis emerge during the more pleasant summer months, and although the hatches may not be as dense as those of the Mother's Day Caddis, they still provide excellent fishing on many Western rivers. Both the emergence and egg laying of these caddis occur en masse, usually in the same parts of the stream and at the same time. Emergence and ovipositing activity may last as long as two hours, and these simultaneous events often trigger a feeding frenzy among the trout. The principle problem for anglers during this hatch is that pupae, emerging adults, and egg-laying females are usually available at the same time, but the trout often choose to feed on one stage over the others. You may have to change patterns repeatedly until you find the right match, though fishing an adult imitation with a pupa or sunken adult as a dropper is an efficient way to test the alternatives.

| | |
|---|---|
| **Other common name:** | Grannom |
| **Family:** | Brachycentridae |
| **Genus:** | *Brachycentrus* |
| **Emergence:** | April–May from midday to late afternoon and July–August in late afternoon to evening; hatches typically last about 2 weeks |
| **Egg-laying flight:** | Afternoon hours in the spring and late afternoon to evening in the summer |
| **Body length:** | ¼–½ inch (6–12 mm) |
| **Hook sizes:** | #12-16, with #14 most common |

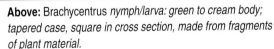

**Above:** Brachycentrus *nymph/larva: green to cream body; tapered case, square in cross section, made from fragments of plant material.*

**Left:** Brachycentrus *pupa: green body with darker markings on top and underside; dark brown to dark gray wing pads.*

**Left:** *Mother's Day Caddis adult: gray wings with brownish cast.*

**Above:** *Male Mother's Day Caddis adult (underside): dark gray to black body; body colors lighter at emergence and darken soon afterward.*

**Left:** *Female Mother's Day Caddis adult (underside): gray to black body. This female was captured while swimming to the river bottom to deposit the bright green egg cluster at the rear tip of the abdomen.*

**Above:** *Black Caddis adult: dark gray to black wings.*

**Right:** *Black Caddis adult (underside): dark gray with olive tones.*

# Important Fishing Stages

**Larva** The larvae of these caddisflies inhabit riffles and runs with moderate to fast flows; they are relatively stationary creatures that remain protected from strong flows that would sweep them away. Trout will eat these cased larvae if they find them tumbling in the current, but in most streams, the number found in the drift is so small that we do not consider the larva a fishable stage of the hatch.

**Pupa** The larvae pupate in same waters in which they live. At emergence, the pupae swim to the surface, drifting downstream as they ascend. In the colder water in spring, the pupae spend more time in the surface film before the adults finally emerge, and trout often hold close to the surface to feed on them. In the warmer summer months, the adults emerge as soon as the pupae reach the surface. Trout take pupae in the surface film, but the window of opportunity is fairly brief, and the fish are more likely to feed on pupae lower in the water column. When large numbers of adults are emerging, plenty of cripples and slow emergers accumulate in the film, and trout often hold near the surface to feed on them. During a hatch, fish pupa patterns close to the surface unless there are few or no visible feeding trout; then fish the pattern close to the bottom, and let it swing it to the surface at the end of the drift.

**Adult** Fishing a dry-fly pattern during this hatch is always effective, especially during the spring. In the colder months, the adults spend more time on the water than they do in warmer weather, and the trout have ample opportunity to feed on them. In fact, getting your fly noticed among the great number of natural insects can pose a problem; targeting specific trout with accurate casts—directly in the feeding lane, close to the fish—is the best recourse.

**Egg-laying female** Generally, as the adults are emerging, egg-laden females are returning to the same water, where they either swim or crawl underwater to lay their eggs on submerged objects. Because of the density and length of these hatches, trout become accustomed to taking adults both on and below the surface, and you can catch fish by imitating either of these stages of the hatch. But during sparser hatches or on heavily fished waters, trout tend to be more selective to the insects subsurface, and your best bet is imitating a swimming female or a sunken, spent female drifting in the current.

# Mother's Day and Black Caddis Pupa Patterns

**BEADHEAD PRINCE NYMPH**

*Originator: Doug Prince*

| | |
|---|---|
| **Hook:** | #12-16 1X–2XL nymph |
| **Head:** | Gold or black metal bead |
| **Weight:** | Lead or nontoxic wire |
| **Thread:** | Black 6/0 or 8/0 |
| **Tail:** | Brown goose or turkey biots |
| **Rib:** | Gold wire |
| **Body:** | Peacock herl |
| **Legs:** | Brown hen-feather barbs |
| **Wing case:** | White goose or turkey biots |

We will be the first to admit that this fly looks nothing like a caddis pupa, and though it is a fine pattern generally, it is particularly effective during *Brachycentrus* hatches—further evidence of the vast gulf between the trout's perception and the angler's. On heavily fished waters, the gold bead may prove too flashy and spook fish. For these occasions, we carry a version tied with a black metal bead or without any bead at all, which is how the original pattern was designed.

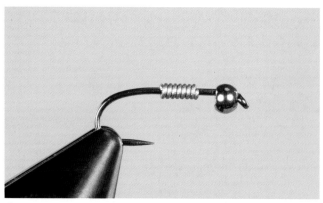

**1.** We prefer tying this pattern, and all bead-head patterns, using countersunk beads, as shown here, since they more easily slide around the hook bend. If possible, choose a wire that, when wrapped around the hook shank, will slide up inside the countersunk hole. This approach reduces bulk on the hook shank and secures the bead tightly. (If using a bead with a uniform-diameter hole, just weight the hook directly behind the bead.) Mount the metal bead by inserting the hook point into the smaller of the two holes; slide the bead up against the hook eye. Take the desired number of wraps of wire, in this case 7 to 8, around the shank, and trim the excess wire.

**2.** Push the wire wraps forward into the countersunk hole until they are tightly seated.

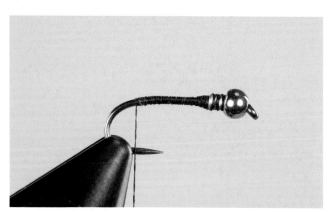

**3.** Mount the thread behind the last wrap of wire. Build a tapered ramp of thread from the hook shank up to the wrapped wire, and secure the wire wraps with thread. Then wrap a thread foundation to the rear of the shank.

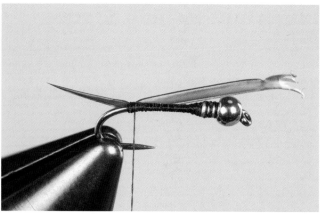

**4.** Clip a biot from a biot strip, and mount it on the far side of the hook shank to form a tail one hook gap in length.

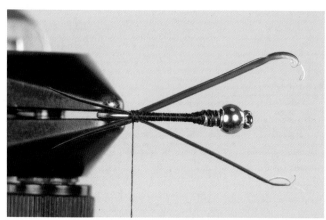

**5.** Mount a second biot on the near side of the shank that matches the first in length, as shown in this top view.

**6.** Clip and bind down the excess biots. Mount a 4-inch length of ribbing wire atop the rearmost tail-mounting wrap. Align the tips of 4 to 6 peacock herls, and trim away the top ¼ inch of the bundle. Then mount the herls and a 6-inch length of scrap tying thread above the rearmost tail-mounting wrap. Position the tying thread behind the metal bead.

**7.** Twist the herl strands and scrap thread together, wrap them up the body to the rear of the bead, and tie them off as shown in the instructions for the Pheasant Tail Nymph, steps 8–9, pp. 22–23.

**8.** Clip and bind the excess herl. Spiral the ribbing wire forward in the opposite direction from the one in which you wrapped the body, taking 4 to 5 wraps to reach the bead. Tie off the wire directly behind the bead, and clip the excess.

**9.** Align the tips of 6 to 8 hen-hackle barbs, and secure them to the far side of the shank to make legs about one hook gap in length. Mount an identical bundle of barbs on the near side of the shank, matching the first bundle in length, as shown in this top view.

**10.** Clip and bind the excess feather barbs. Lay a white biot atop the shank to make about a 15-degree angle with the hook, as shown here from the top. The tip of the biot should extend to the rear of the body. Secure it in position with 2 tight thread wraps; do not clip the excess.

**11.** Mount a second biot to match the first in length and angle. Bind it with 3 tight wraps, but do not clip the excess.

**12.** Fold the butts of the biots rearward, and cover the fold with 3 tight thread wraps, as shown here. This mounting method helps prevent the biots from pulling out. Then clip the excess biots, and bind down the butts. Finish the fly around the thread collar behind the bead.

### STRETCH LACE CADDIS PUPA

**Hook:** #12-14 scud
**Weight (optional):** Lead or nontoxic wire
**Thread:** 6/0 chartreuse and 8/0 dark brown
**Underbody:** Chartreuse thread, colored black or brown with Sharpie permanent marker
**Abdomen:** Medium Chartreuse Ice Stretch Lace
**Wings:** Dark brown Antron yarn
**Wing case:** Brown ⅛-inch Scud Back
**Thorax:** Dark brown Hare-Tron dubbing

The sheen from Ice Stretch Lace is a good imitation of the pupal shuck, so an overbody of Antron or other shiny material is not needed. The two-toned underbody closely matches the body colors of the natural pupa; these underbody colors can be changed to imitate other species of caddis, since caddis pupae are, as a rule, fairly similar in appearance. The weighted version of this pattern, shown in the following sequence, is designed for fishing deeper waters. The unweighted version can be used as a dropper below an adult imitation. To tie the unweighted fly, mount the Ice Stretch Lace ¼ of a hook shank behind the eye; wrap the thread rearward, securing the Stretch Lace to the shank and forming a smooth underbody as you go. Then wrap the material forward, as shown in step 4.

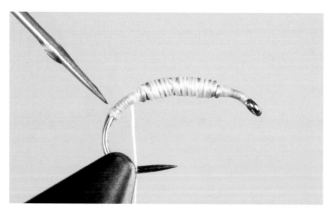

**1.** Form an underbody of wire over the middle of the hook shank, and secure it with the chartreuse thread as described for the Pheasant Tail Nymph, steps 1–2, p. 21. Wrap the thread to the rear of the shank, and form a small thread bump. Position the thread a short distance behind the rearmost wrap of wire, and coat the thread bump with head cement.

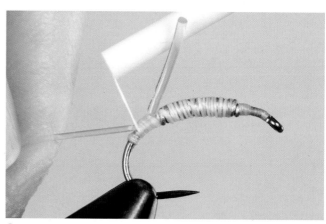

**2.** Mount the strand of body material on top of the hook shank with 4 tight thread wraps, leaving a tag end long enough to grip. Do not clip the tag. Stretch the strand, and wrap the thread rearward, stopping 2 to 3 thread wraps' distance in front of the bump.

**3.** Grip the tag end, stretch it, and cut it close to the hook shank. Flatten the tying thread, and form a smooth underbody over the rear ⅘ of the shank. Tie off the thread and clip it. Mount the brown tying thread behind the eye, and wrap rearward to a point ⅕ of a shank length behind the eye. If you wish, at this point you can coat the thread underbody with head cement and let it dry. The cement will prevent the marker ink from bleeding down the sides of the body and give a more distinct color separation. Color the top of the underbody with the permanent marker.

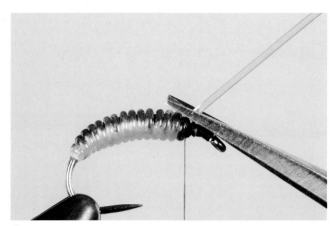

**4.** Stretch the body strand, and place the first wrap directly in front of the bump. Then wrap the strand forward, relaxing the tension slightly with each wrap to form a slightly tapered abdomen up the hanging thread. Use 6 to 8 tight thread wraps to secure the strand. Stretch the strand, and cut it close to the hook shank.

**5.** Cover the butt of the body strand with thread. Position the thread about ⅓ of a shank length behind the hook eye. Mount a 1-inch length of yarn on the near side of the hook shank so that it's angled downward about 45 degrees. Then mount a second length of yarn on the far side of the shank, matching the angle of the first. Trim the excess yarn, and cover the butt ends with the thread. Position the thread at the rearmost yarn-mounting wrap.

**6.** Mount a length of Scud Back on top of the hook shank with 6 to 8 tight thread wraps. Trim the tag end, and position the thread at the rearmost mounting wrap.

**7.** Dub the thorax, stopping 5 to 6 thread wraps' distance behind the eye. Position the thread directly ahead of the thorax.

**8.** Stretch the Scud Back over the top of the thorax, and secure it with 5 to 6 tight thread wraps as shown in this top view.

**9.** Stretch the tag end of the Scud Back, and clip it. Finish the head. Gather the wings, lift them vertically, and trim them at a 45-degree angle so that they are about ¾ the length of the abdomen.

**10.** Preen the wing fibers back into position. Pick out some dubbing fibers on the bottom and sides of the thorax. Trim any excessively long fibers.

**11.** Top view of the finished fly.

**12.** Bottom view of the finished fly.

# Mother's Day and Black Caddis Emerger Patterns

## CDC CADDIS EMERGER

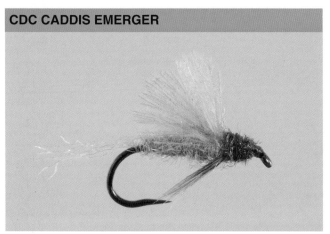

| | |
|---|---|
| **Hook:** | #12-16 standard dry-fly |
| **Thread:** | Brown 8/0 |
| **Tail:** | Tan Antron yarn |
| **Body:** | Green or olive Antron dubbing |
| **Legs:** | Mottled brown partridge or hen-feather barbs |
| **Wing:** | Dark gray or black CDC feather barbs |
| **Head:** | Dark brown dubbing |

This pattern rides quite low and can be difficult to see. In broken water, you may find it easier to track the drift and detect strikes if it's fished as a dropper behind a more visible dry fly. It's a simple fly to dress and worth tying up several, since this sparse pattern with a relatively short CDC wing does not contain a lot of material to float the fly. The wing can become waterlogged or contaminated with algae, in which case it's best to rinse it in clear water, pin it on a patch to dry, and tie on a fresh fly. However, the pattern does fish well subsurface, just beneath the film, again as a dropper behind a dry fly. In a pinch, this fly can be used during a mayfly hatch as well.

**1.** Mount the thread behind the hook eye, and wrap to the rear of the shank. Mount a sparse bundle of Antron fibers to form a tail about one hook gap in length. Clip and bind the excess. Position the thread at the rearmost tail-mounting wrap.

**2.** Dub a slender, slightly tapered body over the rear ²/₃ of the hook shank. Position the thread at the front of the body.

**3.** Strip a bundle of CDC barbs as described for the BWO Puff Fly, steps 1–3, p. 9. Mount the bundle atop the shank, tips pointing rearward, directly ahead of the body. Clip and bind the excess. Position the thread at the rearmost wing-mounting wrap.

**4.** Align the tips of 8 to 10 partridge or hen-feather barbs. Mount them as a bundle on the underside of the shank to form legs that slant rearward to the hook point. You may find this easier if you turn the hook upside down in the vise. Bind and clip the excess.

**5.** Use a thinly dubbed thread to form the head. Finish the fly. Trim the wing fibers to be slightly shorter than the body.

## CDC AND ELK

*Originator: Hans Weilenmann*

| | |
|---|---|
| **Hook:** | #12-16 standard dry-fly |
| **Thread:** | Brown 6/0 |
| **Body/hackle:** | CDC feather |
| **Wing:** | Dark deer hair |

Though the wing on this pattern is actually tied with deer hair, the name of this fly is a tip of the hat to the Elk Hair Caddis, from which it is derived. The palmered CDC feather on this flush-floating pattern does a credible job of imitating the loose nymphal shuck from which the adult insect has almost completely emerged, though the fly can be fished productively just beneath the surface film. The wing gives this fly enough buoyancy to fish very choppy water or for use as a point fly with a weighted dropper. Treat it with powdered floatant only; paste will permanently mat the CDC barbs. The best CDC feather for dressing the fly is one that resembles a partridge body feather in shape, with a relatively short stem and slightly rounded tip.

**1.** Mount the thread about 6 thread wraps' distance ahead of the tailing point on the shank. Wrap rearward to the tailing point. Mount a CDC feather by the tip with 2 tight thread wraps.

**2.** Lift the feather tip, and take one thread wrap directly in front of the tip. Then bind down the feather tip, and wrap a thread foundation to the hook eye. Mounting the feather tip in this fashion prevents it from pulling out when the feather is wrapped.

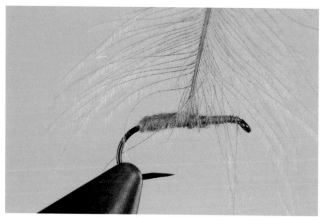

**3.** Clip the butt of the CDC feather in a pair of hackle pliers, and wrap it forward. The material on each wrap should just touch the material from the previous wrap. The rear of the body will be smooth, as shown.

**4.** Continue wrapping forward; some of the CDC barbs will begin to splay, as shown. Tie off the feather about 6 thread wraps' distance behind the hook eye.

**5.** Clip the feather butt and bind down the excess. Position the thread about 6 thread wraps' distance behind the eye. Clean and stack a bundle of deer hair as shown in the instructions for the Sparkle Dun, steps 1–4, pp. 27–28. Size the bundle against the shank, as shown, so that the hair tips extend rearward to the hook bend.

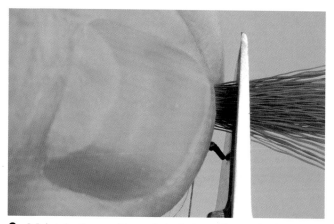

**6.** Maintain the position of the hair, and transfer it to your left fingers. Trim the bundle even with the front of the hook eye.

**7.** Mount the hair atop the shank with 2 to 3 tight thread wraps. Take a third tight wrap at a 45-degree angle through the clipped hair butts, as shown. Take another wrap beneath the hair butts, and finish the fly around the shank beneath the clipped butts.

## Mother's Day and Black Caddis Adult Patterns

**BLACK SNOWSHOE CADDIS**

| | |
|---|---|
| **Hook:** | #12-16 standard dry-fly |
| **Thread:** | Black or dark brown 6/0 or 8/0 |
| **Body hackle:** | Medium dun or dark dun dry-fly |
| **Body:** | Olive dubbing |
| **Wing:** | Black or dark dun snowshoe hare's foot hair |

As is probably obvious, we favor snowshoe hare's foot hair as a winging material because of its excellent flotation and durability; even after catching several trout, the material does not absorb water or break down like deer or elk hair. And it's better suited than deer or elk hair to smaller hook sizes. The dark wing is a good match for the summer hatch of these caddisflies, and this pattern is designed to imitate adult caddis fluttering on the water. The clipped hackle places the body closer to the surface and helps the fly land upright, while the wing provides enough buoyancy to float the fly even in rough water or when supporting a lightly weighted dropper pattern.

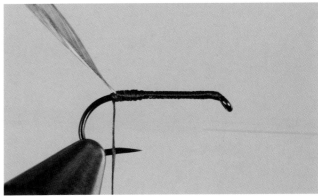

**1.** Mount the thread behind the hook eye, and wrap a thread foundation, stopping about 3 thread wraps' distance ahead of the normal tailing point on the hook. Prepare and mount a hackle as described for the Little Green Drake, steps 4–6, p. 42. When mounting the feather, leave a length of bare hackle stem between the rearmost mounting wrap and the lowermost barbs on the feather. The length of this bare stem should be equal to ½ the thickness of the dubbed abdomen. Position the thread just behind the hackle.

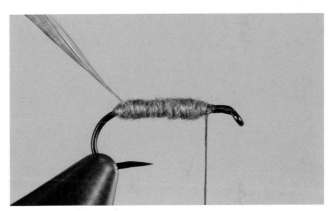

**2.** Beginning behind the mounted hackle, dub the abdomen over the rear ¾ of the shank. Position the thread directly ahead of the abdomen.

**3.** Wrap and tie off the hackle as shown in the instructions for the Deer Hair Caddis, steps 3–5, p. 132. Position the thread 3 thread wraps' distance in front of the body.

**4.** Prepare a bundle of snowshoe hare's foot hair as shown in the instructions for the BWO Snowshoe Dun, steps 1–2, p. 13. Mount the hair as described for the PMD Hare-Wing Dun, steps 6–7, p. 32, and then finish the fly and trim the hair as shown in step 8.

**5.** Trim the hackle barbs beneath the shank so that they are even with the hook point.

## CALM-WATER CADDIS

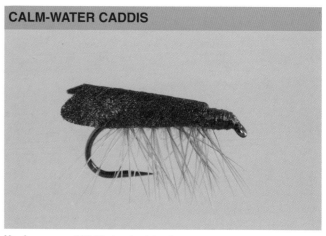

| | |
|---|---|
| **Hook:** | #12-16 standard dry-fly |
| **Thread:** | Dark brown 8/0 |
| **Body hackle:** | Medium dun or dark dun dry-fly |
| **Body:** | Dark olive dubbing |
| **Wing:** | Black or dark dun Thick Wing |

As the name suggests, this somewhat more realistic caddis pattern is intended for fishing smoother water, where selective trout get a good, long look at anything passing by. This fly does not float particularly well; it rides low and is easily swamped in choppier water. Even on a calmer surface, it's important to treat the fly with a liquid or powdered floatant, since the pattern contains relatively little material that contributes to flotation. If there's a trick in dressing this pattern, it's keeping the wing mounted low over the body; to that end, the hackle is trimmed flush on top. If the wing silhouette is too high, the fly will spin during casting and twist a light tippet. This is a good fly for tough fish in tough water. We prefer Thick Wing for a winging material because of its durability, but shaping it requires a wing cutter; trimming the material with scissors is difficult and gives poor results. You can, however, form the wing from other materials that are more easily trimmed with scissors, such as the dark dun Medallion Sheeting shown in steps 2–3.

**1.** A quick and consistent way to form the wing is to use a wing cutter. The one shown here is a Caddis Wing Cutter from River Road Creations. The cutters come in sizes 8 to 18 and are accurately sized for folded tent wings.

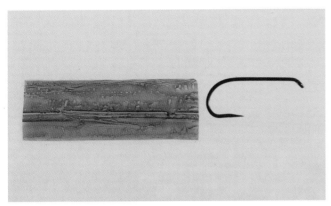

**2.** To form the wing from another sheet material, such as the dark dun Medallion Sheeting shown here, cut a strip 1½ times the width of the hook gap. Then cut a piece 1 inch long from the strip.

**3.** Fold the strip lengthwise as shown at the top. Trim the edge opposite the fold to the shape shown in the middle. The unfolded finished wing should resemble the shape shown at the bottom. The finished wing should be about twice as long as the hook shank.

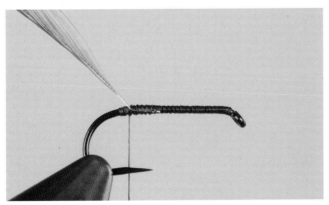

**4.** Mount the thread behind the hook eye, and wrap a thread foundation, stopping about 3 thread wraps' distance ahead of the normal tailing point on the hook. Prepare and mount a hackle as described for the Little Green Drake, steps 4–6, p. 42. When mounting the feather, leave a length of bare hackle stem between the rearmost mounting wrap and the lowermost barbs on the feather. The length of this bare stem is equal to ½ the thickness of the dubbed abdomen. Position the thread just behind the hackle.

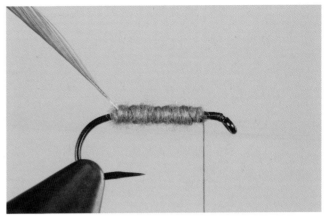

**5.** Beginning behind the mounted hackle, dub the abdomen over the rear ¾ of the shank. Position the thread directly ahead of the abdomen.

**6.** Wrap and tie off the hackle as shown in the instructions for the Deer Hair Caddis, steps 3–5, p. 132. On this fly, however, most of the hackle barbs will be trimmed away. To compensate for this loss, begin wrapping the hackle by taking 2 closely spaced wraps at the rear of the body; then spiral the hackle forward over the abdomen, and take 2 closely spaced wraps at the front of the abdomen. Clip and bind the excess hackle.

**7.** Trim the barbs above the hook shank, flush with the top of the body. Then clip a V in the hackle on the underside of the shank, removing the barbs from an arc of about 90 degrees centered under the hook shank, as shown in this front view.

**8.** To get the wing to lie flat over the body, form a thread base ahead of the body that is the same thickness as the front of the body. Position the thread at the front edge of the body.

**9.** Center the wing over the top of the hook shank with the rear of the wing aligned with hook bend. Secure the wing with 3 thread wraps. Release the wing, and check to make sure it's centered over the hook shank, as shown in this top view. If the wing is crooked, remove one thread wrap, and gently move the wing into position.

**10.** Once the wing is positioned correctly, secure it with additional thread wraps. Trim and bind the excess. Form the head from tying thread, then finish the fly.

# Mother's Day and Black Caddis Egg-Laying Patterns

**MOTHER'S DAY CADDIS**

*Originator: Doug Duvall*

| | |
|---|---|
| **Hook:** | #12-16 1XL nymph |
| **Thread:** | Black 8/0 |
| **Abdomen:** | Peacock herl |
| **Wing:** | Brown partridge-feather barbs |
| **Thorax:** | Dark gray dubbing |

We'll certainly admit that this fly doesn't look like all that much, but as with many simple patterns, this one is designed to appeal to the fish rather than the angler. All the materials in this pattern readily absorb water, so the fly sinks well. Use it as a dropper behind a dry fly, or below an indicator with split shot on the leader in deeper or faster water. Fish it drag-free, and then allow the fly to swing on a tight line at the end of the drift so that it rises in the water. This is a productive sunken pattern for the spring hatch; for the summer Black Caddis hatch, use dark dun or black feather barbs for the wing.

**1.** Mount the thread behind the hook eye, and wrap it to the rear of the shank. Align the tips of 4 to 6 peacock herls, and trim away the top ¼ inch of the bundle. Then mount the herls and a 6-inch length of scrap tying thread atop the rearmost tail-mounting wrap. Position the tying thread ⅓ of a shank length behind the eye.

**2.** Twist the herl strands and scrap thread together to form a chenille-like strand. Wrap the strand over the rear ²/₃ of the shank to form the abdomen, pausing as needed to retwist the strand. Tie off the herl at the hanging thread.

**3.** Clip and bind down the excess herl, and position the thread at the front of the abdomen. Preen the lower barbs of a partridge feather toward the butt end of the feather, leaving a tip section about 1½ shank lengths long.

**4.** Position the feather flat atop the shank with the front side of the feather facing upward. Mount the feather to make a wing that extends just beyond the rear of the hook.

**5.** Clip and bind the excess wing material, and position the thread at the rearmost wing-mounting wrap. Dub the thorax, and finish the fly.

### CDC SPENT CADDIS

| | |
|---|---|
| **Hook:** | #14-18 standard nymph |
| **Thread:** | Black 8/0 |
| **Egg sac (optional):** | Bright green Antron dubbing |
| **Body:** | Dark olive dubbing |
| **Wing:** | Dark dun CDC barbs |

Though CDC is typically used on dry flies, it makes an excellent material for subsurface patterns as well. On this sunken female caddis imitation, the CDC barbs are quite mobile in the water, simulating the active movements of the diving insect, and the fibers capture tiny air bubbles that shimmer like the wings of the natural. Since the CDC wants to float, gently squeeze water into the wing with your fingertips until the barbs are saturated. Fish the fly as a dropper about 8 to 10 inches behind a dry fly or small indicator. If you want this fly to sink more quickly, you can add lead or nontoxic wire to the hook shank or a split shot to the tippet. You can tie a floating version of this fly by using a dry-fly hook instead of a nymph hook and forming the wing from snowshoe hare's foot hair, which resists waterlogging better than CDC and is more durable. The tying steps are exactly the same as those shown for the Black Snowshoe Caddis, pp. 148–149, except that the hackle is omitted; the wings are then divided as explained in steps 5–11 in the following sequence. Fish this version drag-free on the surface; at the end of the drift, pull the fly underwater and swim it down and across, just beneath the surface, to imitate a swimming adult female.

**1.** Mount the thread behind the hook eye, and wrap a foundation to the rear of the shank. If an egg cluster is desired, apply the green dubbing over the rear ¼ of the hook shank.

**2.** Dub the body to within 6 to 8 thread wraps' distance behind the hook eye. Position the thread 2 to 3 thread wraps' distance ahead of the body.

**3.** Strip a bundle of barbs from one or two CDC feathers as described for the BWO Puff Fly, steps 1–3, p. 9. The barbs should be longer than the hook shank. Position the bundle of barbs atop the shank so that the tips extend beyond the hook bend.

**4.** Mount the barbs atop the shank with 3 tight thread wraps.

**5.** With your fingers or a dubbing needle, divide the CDC barbs into 2 equal bundles, as shown in this top view.

**6.** Take 2 or 3 crisscross wraps between the wings to lock the barb bundles in position, as shown in this top view.

**7.** Hold the wings against the sides of the body, and take a thread wrap or two over the rear of the crisscross wraps, toward the body.

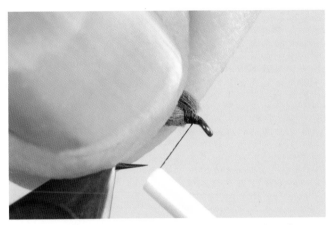

**8.** Lift the butts of the CDC barbs, and take 2 thread wraps against the base of the barbs.

**9.** Tie off and cut the thread. Trim the CDC butts even with the hook eye. Gather the wings together, draw them rearward, and trim them so they extend about ½ a hook gap beyond the hook bend.

**10.** Here's a top view of the finished fly.

**11.** The wings should extend outward from the sides of the body, as shown in this front view.

# Small Black Caddis

The larvae of these small caddisflies inhabit the weedy sections of streams that have moderate current flows; tailwaters and spring creeks with abundant aquatic vegetation can produce impressively large populations of these insects. Depending on the particular stream, the seasonal hatch cycle may last from two weeks to two months, but regardless of the duration, the density of hatching insects usually peaks during the middle of the cycle. Daily emergences typically last for one to two hours. Unlike other caddis species, whose emergences may consist of only a modest number of flies hatching at any one time, the Small Black Caddis emerge en masse. And, in fact, the number of insects available to the trout during a hatch is increased substantially because recently hatched females often return to the stream to lay their eggs as other adults are just emerging. So what these small caddis may lack in size, they make up for in density on the water.

| | |
|---|---|
| **Other common name:** | Little Western Weedy-Water Sedge |
| **Family:** | Brachycentridae |
| **Genus:** | *Amiocentrus* |
| **Emergence:** | June to October, midday to late afternoon |
| **Egg-laying flight:** | Midday to late afternoon |
| **Body length:** | ¼–5⁄16 inch (6–8 mm) |
| **Hook sizes:** | #16-18 |

**Above:** *Small Black Caddis larva: green body; tapered case, round in cross section, made from plant material.*

**Above:** *Small Black Caddis pupa: green body with darker markings on top and underside; dark brown to black wing pads.*

**Left:** *Small Black Caddis adult: dark gray to black wings.*

**Above:** *Small Black Caddis adult (underside): dark gray with olive tones.*

# Important Fishing Stages

**Larva** Small Black Caddis larvae move about to feed and are occasionally swept away by the current, where they are available to trout. But the number of insects found in the drift at any given time is not really consequential, and we don't regard the larvae of this caddis to be a fishable stage of the life cycle.

**Pupa/emerger** When the daily hatch begins, large numbers of pupae emerge from cases attached to the streambed and make their way upward through the water. They are not notably strong swimmers, generally drifting a long distance with the current before they reach the surface, so they are vulnerable to predation for a significant interval of time. Trout feed on these pupae throughout the water column, and a pupa imitation fished deep below an indicator—either as a dropper behind a larger, weighted nymph or with split shot on the leader—will most likely take some fish. But more typically, the trout tend to take up stations closer to the surface, where they expend less energy feeding on pupae hanging in the surface film. These fish can be targeted with a pupa or emerger pattern fished behind a dry fly or small indicator.

**Adult** Because of their small size and dark color, the Small Black Caddis adults can be difficult to see on the water unless they're hatching in significant numbers. If you don't see trout rising to these insects, fish a pupa pattern rather than an adult imitation to target the subsurface feeders. Once the adults emerge, they spend very little time on the surface before flying off, and when trout key in on the newly hatched adults, they often taken them in splashy, aggressive rises in order to capture the insects before they escape.

**Egg-laying female** Some of the females may deposit their eggs on the surface, but most of them swim or crawl underwater and lay their sticky green egg masses on submerged objects. After their task is complete, a few females may manage to return to the surface, but the vast majority are caught in the current and swept downstream. Trout feed on egg-laying females all through the water column, so an egg-laying pattern can be presented at any depth. But some of the best fishing occurs in eddies and current seams, where spent females accumulate and make an easy meal for the fish.

# Small Black Caddis Pupa Patterns

**KRYSTAL FLASH PUPA**

*Originator: Gary LaFontaine*

| | |
|---|---|
| **Hook:** | #16 1XL nymph |
| **Thread:** | Black 8/0 |
| **Overbody:** | Black Antron yarn |
| **Body:** | Peacock or green Krystal Flash |
| **Legs:** | Dark brown mottled hen-hackle barbs |
| **Head:** | Dark brown dubbing |

While Gary LaFontaine's Sparkle Pupa design is adaptable to many caddis species, this version imitating the Small Black Caddis is a particularly good producer. Year in and year out, it is arguably our top pattern for this hatch. With the non-absorbent body and shuck materials and the relatively heavy hook wire, the fly easily penetrates the surface film but is still light enough to be fished as a dropper behind the small dry flies used to imitate the adults. It can also be fished deeper in the water column by using it as a trailing fly on a tandem nymph rig or adding split shot to the leader. When tying the fly, form a relatively sparse Antron capsule; not much material is needed on a small hook.

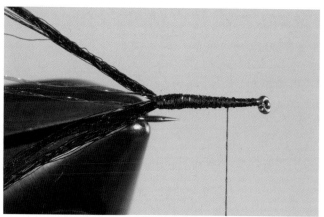

**1.** Mount the thread behind the hook eye, and lay a thread foundation to the rear of the shank. Prepare 2 strands of Antron yarn as described for the Deep Sparkle Pupa, step 1, p. 126. Mount them on an unweighted hook as shown in steps 2–3. Clip and bind the excess. Mount 3 strands of Krystal Flash atop the rearmost thread wraps used to mount the Antron yarn. Clip and bind the excess. Position the tying thread ⅓ of a shank length behind the eye, as shown in this top view.

**2.** Gather the strands of Krystal Flash into a bundle, and twist them clockwise (when viewed from above). Wrap the twisted strand forward to form a body over the rear ⅔ of the shank. When you reach the tying thread, secure the Krystal Flash.

**3.** Clip the excess Krystal Flash. Form the Antron shuck as described for the Deep Sparkle Pupa, steps 5–8, p. 127.

**4.** Clip and bind down the excess yarn, and position the thread at the front of the Antron bubble. Align the tips of 6 to 8 hen-hackle barbs. Mount them underneath the shank to form legs about as long as the body. You may find it easier to invert the hook to mount the legs.

**5.** Clip and bind the excess feather barbs. Use a thinly dubbed thread to form the head, and finish the fly.

## SUPER PUPA

| Hook: | #16-18 1XL nymph |
|---|---|
| Thread: | Chartreuse 6/0 and dark brown 8/0 |
| Underbody: | Chartreuse thread |
| Overbody: | Green or chartreuse Super Hair |
| Wing buds: | Black Antron yarn |
| Head: | Dark brown dubbing |

Chartreuse thread is used for the underbody on this pattern because the Super Hair overbody is translucent; a darker underbody thread will mute the color. The chartreuse materials here may seem oddly bright, but they are a reasonably good match for the body color of the natural. Maintaining a flattened thread when you dress the underbody will produce a smooth foundation for a neatly formed overbody. The Super Hair gives a segmented appearance to the body and a shimmer that is suggestive of the pupal shuck of the natural insect; the glossiness is enhanced by coating the body with head cement, which also increases durability. The smooth body material allows this pattern to sink quickly, though the fly is still sufficiently light to be used as a dropper behind an adult imitation.

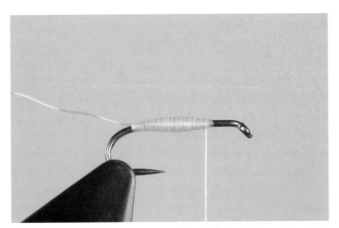

**1.** Mount the chartreuse thread ⅓ of a shank length behind the hook eye, and wrap it smoothly to the rear of the shank. Mount one strand of Super Hair at the rear of the shank. Flatten the thread by spinning the bobbin counterclockwise (when viewed from above), and form a tapered thread underbody over the rear ⅔ of the hook shank. Position the thread at the front of the underbody.

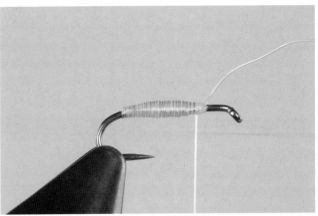

**2.** Wrap the Super Hair strand over the underbody; place each wrap snugly against the previous one. Secure the strand at the hanging thread.

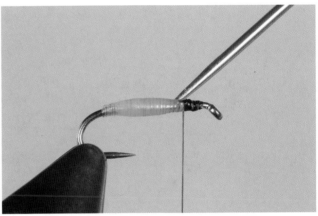

**3.** Clip the excess body material, and tie off the underbody thread. Mount the brown thread behind the hook eye, and position it at the front of the body. Lightly coat the body and thread wraps with head cement.

**4.** Cut a 2-inch length of Antron yarn. Center it against the near side of the hook shank, so that one end of the yarn is angled down and to the rear at 45 degrees. Secure the yarn in this position with 3 thread wraps.

**5.** Fold the forward-facing tag of the yarn over the top of hook, and position it against the far side of the shank, pointing downward and rearward at a 45-degree angle. Secure it with 3 tight thread wraps directly over the previous wraps, as shown in this top view.

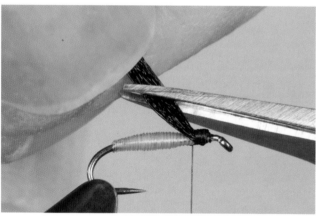

**6.** Gather the yarn fibers from both sides of the hook, and pull them upward. Trim the yarn parallel to the hook to form wings ½ to ⅔ the length of the hook shank.

**7.** Preen the wing fibers back into position. Apply a small amount of dubbing to the thread, and form the head. Then finish the fly.

# Small Black Caddis Emerger Patterns

## SUPER CADDIS EMERGER

| | |
|---|---|
| **Hook:** | #16-18 standard dry-fly |
| **Thread:** | Chartreuse 6/0 and dark brown 8/0 |
| **Tail:** | Tan Antron yarn fibers |
| **Underbody:** | Chartreuse thread |
| **Overbody:** | Green or chartreuse Super Hair |
| **Wing:** | Black or dark dun CDC feather barbs |

With a smooth, quick-sinking body that pulls the fly downward and a buoyant CDC wing that holds it up, this pattern rides tight to the film—just above it if the wing is treated with a powder floatant, just below it if the wing is left untreated and saturated with water before fishing. This fly is designed to be fished as a dropper behind an imitation of the adult, but it also makes a good egg-laying female pattern that can be fished anywhere in the water column. Forming a smooth thread underbody is the key to tying a neatly segmented body.

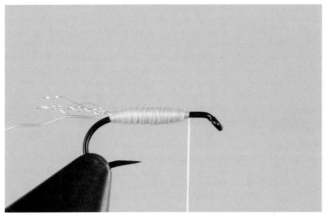

**1.** Mount the chartreuse thread ⅓ of a shank length behind the hook eye, and wrap it smoothly to the rear of the shank. Mount one strand of Super Hair at the rear of the shank. Mount a sparse bundle of Antron yarn fibers to make a tail one hook gap in length. Clip and bind the excess. Flatten the thread by spinning the bobbin counterclockwise (when viewed from above), and form a tapered thread underbody over the rear ⅔ of the hook shank. Position the thread at the front of the underbody.

**2.** Wrap the Super Hair strand over the underbody, and secure the strand at the hanging thread.

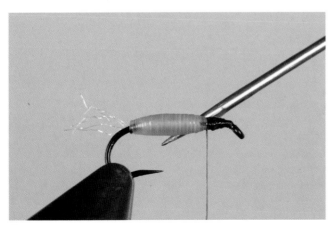

**3.** Tie off the chartreuse thread. Mount the brown thread behind the hook eye, and position it at the front of the body. Lightly coat the body and thread wraps with head cement.

**4.** Strip the barbs from 1 to 3 CDC feathers, as described for the BWO Puff Fly, steps 1–3, p. 9. Position the bundle over the hanging thread with the tips extending rearward past the hook bend. Secure the bundle atop the shank with 4 tight thread wraps.

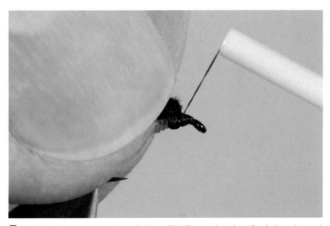

**5.** Lift the butt ends of the CDC, and take 2 tight thread wraps against the front base of the bundle.

**6.** Finish the head around the hook shank, beneath the CDC butts, and clip the thread. Trim the CDC butts, following the angle of the hook eye, to leave a tuft. Trim the wing even with the hook bend.

## BLACK SNOWSHOE EMERGENT CADDIS

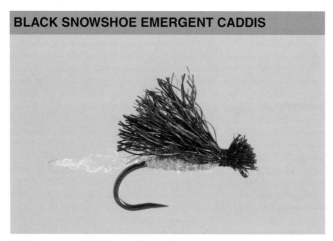

| Hook: | #16-18 standard dry-fly |
| Thread: | Black 8/0 |
| Tail: | Tan Antron yarn |
| Body: | Light green Antron dubbing |
| Wing: | Black or dark dun snowshoe hare's foot hair |

This flush-floating pattern represents an adult caddis, fully emerged, with the pupal shuck still attached. Because the Antron shuck fibers are fused together at the end, they remain consolidated into a hollow, shell-like form that imitates the shuck better than would loose, splayed fibers. The key to forming the shuck is using a sparse bundle of Antron fibers; use too much material, and you may burn your fingertips when the fibers are melted. This fly is quite durable, floats well, and can be treated with either a paste or powder floatant.

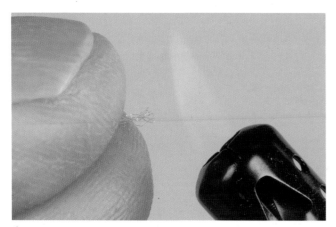

**1.** To form the pupal shuck, pinch a sparse bundle of Antron fibers between thumb and forefinger. Trim the end of the bundle so that a ⅛-inch stub of fibers extends beyond your fingertips. Briefly touch the flame from a butane lighter to the stub of yarn so that the fibers melt back to your fingertips.

**2.** As soon as the fibers melt, roll the tips of your fingers forward and lightly compress the melted tips together.

**3.** The ends of the fibers will fuse together to form the rear of the shuck.

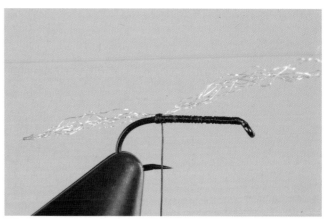

**4.** Mount the thread behind the hook eye, and wrap a foundation to the rear of the shank. Using firm but not excessively tight thread wraps, mount the Antron shuck atop the shank so that it extends rearward the length of the hook shank. Do not clip the excess.

**5.** Gently pull the tag ends of the yarn fibers forward; this will cause the shuck to form a slender bubble. If the bubble is too large, gently pull the melted end rearward to make it smaller.

**6.** After the bubble is formed, secure the yarn butts with 3 tight thread wraps, and trim the excess. (After trimming the excess Antron fibers, repeat steps 1–2 on the leftover yarn so that it is ready to form a shuck on the next fly.) Dub a tapered body over the rear ¾ of the hook shank. Position the thread in front of the body.

**7.** Prepare a bundle of snowshoe hare's foot hair as described for the BWO Snowshoe Dun, steps 1–2, p. 13. Position the bundle atop the hook shank with the tips extending past the hook bend. Secure the bundle in front of the body with 4 or 5 tight thread wraps.

**8.** Lift the hair butts, and take 2 tight thread wraps against the front base of the bundle. Finish the fly around the shank, beneath the hair butts.

**9.** Trim the butt ends of the hair, following the angle of the hook eye, to leave a tuftlike head. Then trim the wing at an angle so that it extends to the bend of the hook.

# Small Black Caddis Adult Patterns

## SMALL BLACK SNOWSHOE CADDIS

**Hook:** #16-18 standard dry-fly
**Thread:** Black 8/0
**Body hackle:** Dark dun dry-fly
**Body:** Fine light olive dubbing
**Wing:** Black or dark dun snowshoe hare's foot hair

This fly is a productive imitation of a fluttering adult caddis. The palmered hackle and snowshoe hare's foot wing make this a highly buoyant fly that can be fished in broken water, and even in size 18 it floats well enough to support a pupa or sunken adult pattern on a dropper. While the dark snowshoe hare imitates the wing color of the natural insect, it has the added benefit of making this small fly highly visible in overcast or low-light conditions. Odd as it may seem, this pattern is also an effective imitation of a midge cluster. Except for the hook size and the colors of the materials, this fly is identical to the Black Snowshoe Caddis pattern used to imitate the Mother's Day Caddis adult and is dressed using the instructions on pp. 148–149.

## ANTRON CALM-WATER CADDIS

**Hook:** #16-18 standard dry-fly
**Thread:** Black 8/0
**Body hackle:** Dark dun dry-fly
**Body:** Fine light olive dubbing
**Wing:** Black Antron yarn

Though this pattern is similar in appearance to the Small Black Snowshoe Caddis shown above, the palmered hackle is trimmed on the bottom so that the fly sits down nearly flush on the surface, and the Antron wing gives a little trimmer profile than the snowshoe hare's foot. As a result, this pattern is better suited to smooth, glassy currents, where more selective trout often feed and where superior flotation in a fly is not normally required. Treat the fly with a powder or liquid floatant. The enlarged thread foundation ahead of the body is important for keeping the wing low over the back of the fly, and the wing should not be too densely dressed. The body of this pattern is identical to the one shown on the Calm-Water Caddis used for the Mother's Day Caddis hatch, so the following instructions are abbreviated.

**1.** Dress the body, hackle the fly, clip the hackle (shown here from the front), and wrap the wing-mounting foundation as described for the Calm-Water Caddis, steps 4–8, pp. 150–151.

**2.** Cut a 2-inch length of Antron yarn, and split it lengthwise to get a winging strand of the proper thickness for the hook size. Mount the strand ahead of the body with 4 to 5 tight thread wraps so that it lies flat over the top of the fly and extends about a shank length beyond the hook bend. Lift the front tag of the yarn, and take 2 wraps tightly abutting the front of the bundle.

**3.** Finish the fly around the hook ahead of the yarn tag. Trim the front tuft of yarn, following the angle of the hook eye, to form a tuftlike head. Trim the rear of the wing at an angle so that it's even with the hook bend.

**4.** Here's a top view of the finished fly.

# Small Black Caddis Egg-Laying Patterns

## CDC SMALL BLACK CADDIS

| | |
|---|---|
| **Hook:** | #16-18 standard dry-fly |
| **Thread:** | Black 8/0 |
| **Body:** | Fine dark green or black dubbing |
| **Wing:** | Black or dark dun CDC barbs |

This basic design ties up quickly and easily, and it's our pattern of choice for imitating egg-laying and diving caddis. It's equally effective treated with powder floatant and fished on the film or saturated with water and fished subsurface, where the CDC gives good movement to the fly. When trout are feeding on ovipositing Small Black Caddis, you can use a dropper rig to fish a pair of these flies in tandem, one on the surface and one below.

**1.** Mount the thread behind the hook eye, and wrap a foundation to the rear of the shank. Dub a body forward over the rear $4/5$ of the hook shank. Position the thread at the front of the body.

**2.** Prepare a bundle of CDC barbs as described for the BWO Puff Fly, steps 1–3, p. 9. Mount them atop the shank, tips pointing rearward, directly ahead of the body, with 3 to 4 tight thread wraps. Lift the butt ends of the barbs at the front of fly, and take 2 tight thread wraps against the front base of the barbs. Finish the head of the fly around the hook shank beneath the CDC butts.

**3.** Trim the CDC butts even with the hook eye to form a tuft-like head. Trim the wing fibers even with the hook bend.

## KRYSTAL FLASH SOFT-HACKLE

| | |
|---|---|
| **Hook:** | #16-18 1XL nymph |
| **Thread:** | Black 8/0 |
| **Abdomen:** | Black Krystal Flash |
| **Thorax:** | Fine dark brown dubbing |
| **Hackle:** | Black hen or starling |

The soft-hackle style of fly is a natural for the caddis hatch, particularly for imitating subsurface egg layers. The fly can be fished dead-drift to imitate a spent female tumbling in the current; it can be fished with movement you impart to imitate a swimming caddis; or it can be fished drag-free and allowed to swing at the end of the drift to imitate both subsurface behaviors of the natural insect. We dress this pattern with a bit more hackle than is customary for soft-hackle designs; the hackle feather is left intact, rather than stripped on one side.

**1.** Prepare the hackle by stripping away barbs from the butt end of the feather until the lowermost barbs on the stem are about as long as the hook shank. Mount the hackle feather as shown in the instructions for the Partridge and Green, steps 2–3, p. 128. Instead of mounting the floss shown in those steps, however, mount 2 strands of Krystal Flash, securing them atop the hook as you wrap rearward to the end of the shank.

**2.** Flatten the thread by spinning the bobbin counterclockwise (as viewed from above), and wrap forward, forming a smooth underbody over the rear ²/₃ of the shank.

**3.** Draw the strands of Krystal Flash into a bundle. Twist the bundle clockwise (when viewed from above), and wrap a body over the rear ²/₃ of the shank. Pause as needed to retwist the Krystal Flash. Tie off the Krystal Flash when you reach the hanging thread.

**4.** Clip the excess Krystal Flash. Dub a thorax to the hackle tie-in point. Position the thread at the front of the thorax.

**5.** Wrap and tie off the hackle as shown in the instructions for the Partridge and Green, steps 7–8, p. 129. Trim the excess. Form a small thread head in front of the frontmost hackle wrap, and finish the fly.

# Spotted Sedge

These widely distributed caddisflies are found in most Western trout streams, though their numbers vary significantly depending on the character of the water. Unlike many other caddis species, the Spotted Sedge does not build a case that it can carry around as it moves. Instead, the larva fashions a fixed shelter on rocks or streambed debris in riffles and shallow runs, and then spins a net with which it seines tiny organisms from the current. Thus the largest populations of these caddisflies usually occur in plankton-rich waters downstream of lakes and dams, where the food supply is abundant; freestone streams fed by runoff, on the other hand, have far less food in the drift and thus smaller numbers of Spotted Sedge. With a long seasonal emergence period, these caddis are a staple for both trout and anglers on many streams, and they are often responsible as well for one of the classic scenarios in fly fishing—the evening hatch. Both the emerging adults and egg-laying females appear on the water simultaneously, en masse, typically just after sunset and provide some challenging fishing in low-light conditions.

| | |
|---|---|
| **Other common names:** | None |
| **Family:** | Hydropsychidae |
| **Genus:** | *Hydropsyche* |
| **Emergence:** | May to October, with peak periods in June–July and September, during the evening hours |
| **Egg-laying flight:** | Late afternoon or evening |
| **Body length:** | ⅜–⁹⁄₁₆ inch (10–14 mm) |
| **Hook sizes:** | #12-14, with #14 most common |

**Above:** *Spotted Sedge nymph/larva: olive or tan to brown on top; underside is lighter shade of green or tan.*

**Above:** *Spotted Sedge pupa: tan to brown or olive body, often with shades of green; medium brown to dark brown wing pads.*

**Left:** *Spotted Sedge crippled emerger: the crumpled wings are a conspicuous feature of this crippled emerging adult. Both the bright body color and light wing color darken shortly after emergence.*

**Above:** *Spotted Sedge crippled emerger (underside): the bubbles trapped inside the translucent pupal shuck are clearly visible.*

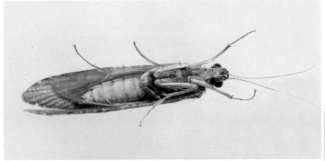

**Top left:** *Spotted Sedge adult (underside): some adults have tan or brown bodies.*

**Top right:** *Spotted Sedge adult (underside): other adults have olive or green bodies.*

**Left:** *Spotted Sedge adult: brown wings, often spotted with gray.*

# Important Fishing Stages

**Larva** Even though these larvae live in crude shelters from which they tend their nets, good numbers of them are found drifting in the current in the morning and evening hours. At maturity, the larvae enclose themselves in their shelters to pupate for about two to three weeks, during which time they are unavailable to the trout. The best time to fish a larva pattern, then, is before the larvae pupate, about three to six weeks before the start of a hatch cycle. The larvae are most often found drifting close to the streambed, and a larva imitation is most productive when fished just off the bottom.

**Pupa/emerger** At emergence the pupae leave their shelters en masse, drifting downstream as they swim to the surface, where they emerge from their pupal shucks and fly off. Trout take these pupae anywhere in the water column, and a pupa pattern can be fished at any depth. But trout often focus most keenly on pupae close to the surface, where the insects are concentrated. The daily hatch period is not unusually long, perhaps an hour or so, but large numbers of these caddis emerging can spark a feeding frenzy among the trout.

**Adult** After leaving their pupal shucks, the adults spend little time on the surface and fly off rather quickly, so trout tend to key on the pupae more than the adults. During dense hatches, however, crippled emergers and lagging adults are always present on the water, and trout readily take this easy prey. Fishing a pupa pattern as a dropper behind an adult imitation is a particularly effective way to fish this hatch.

**Egg-laying female** Egg-laden females deposit their eggs on submerged rocks in riffles and runs either by landing on the surface and swimming down to the streambed or by landing on exposed rocks and then crawling underwater. Once egg laying is complete, the females attempt to return to the surface. A great many are simply caught in the current and swept downstream; some do manage to make it to the surface, but few become airborne again. A soft-hackle pattern is a good imitation of a swimming or spent submerged female; a hackleless dry-fly pattern works well for a floating spent female.

# Spotted Sedge Larva Patterns

**WIRED CADDIS LARVA**

| | |
|---|---|
| **Hook:** | #12-14 scud |
| **Head:** | Black metal bead |
| **Weight:** | Lead or nontoxic wire |
| **Thread:** | Black 6/0 or 8/0 |
| **Abdomen:** | Green or olive wire |
| **Wing case:** | Dark olive or brown ⅛-inch Scud Back |
| **Thorax:** | Brown ostrich herl |

With a weighted metal bead head, wire abdomen, absorbent herl thorax, and streamlined shape, this fly sinks quickly—an important attribute, since a Spotted Sedge larva pattern should be fished close to the bottom. When these larvae are present in the drift, trout often move into very shallow water to feed on them. Fishing this pattern as a dropper behind a dry fly is a good way to target these fish. In faster or deeper water, fishing the fly below an indicator with split shot added to the leader may be necessary to get it to a productive level in the water column.

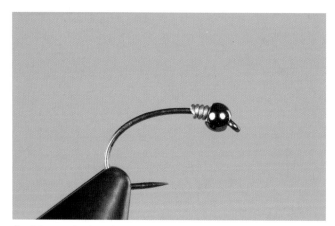

**1.** Mount the bead and 6 wraps of wire as described for the Beadhead Prince Nymph, steps 1–2, p. 141.

**2.** Mount the thread behind the rearmost wrap of wire, and secure the wire with thread wraps. Mount a 6-inch length of the body wire atop the shank behind the wrapped wire. Secure the body wire by the very end, leaving no tag. Bind the body wire atop the shank, wrapping to the rear of the hook. Flatten the thread by spinning the bobbin counterclockwise (as viewed from above), and wrap a slightly tapered underbody over the rear ⅔ of the shank. Position the thread ⅓ of a shank length behind the hook eye.

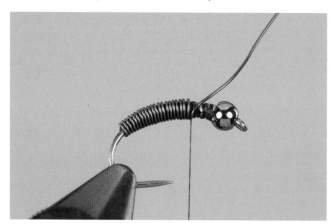

**3.** Wrap the wire forward, and secure it at the hanging thread.

**4.** Clip and bind down the excess wire. Mount a strip of Scud Back atop the shank, directly in front of the abdomen. Clip and bind the excess. Mount an ostrich herl by the tip atop the rearmost thread wrap securing the Scud Back. Clip the excess, and wrap the thread forward, forming a smooth foundation to a point about 4 thread wraps' distance behind the bead.

**5.** Wrap the herl forward to the tying thread in close but not overlapping wraps. Secure the herl at the hanging thread.

**6.** Clip the excess herl. Draw the Scud Back over the top of the herl, and secure it directly behind the bead, as shown in this top view. Clip and bind the excess, and finish the fly behind the bead.

## STRETCH LACE LARVA

| | |
|---|---|
| **Hook:** | #12-16 2XL–3XL curved nymph |
| **Weight:** | Lead or nontoxic wire |
| **Thread:** | White and brown 6/0 or 8/0 |
| **Underbody:** | White thread |
| **Abdomen:** | Olive, green, or tan Ice Stretch Lace |
| **Wing case:** | Brown ⅛-inch Scud Back |
| **Thorax rib:** | Dark brown 3/0 or 6/0 thread |
| **Thorax:** | Dark olive-brown dubbing |

A long-shank hook gives this Spotted Sedge larva imitation the proportions of the natural insect. The elastic cord used to dress the abdomen creates realistic body segmentation and promotes fast sinking. Larva color can vary somewhat from stream to stream; choose a body material color that most closely matches that of the natural insect on the waters you intend to fish. Since the body material is translucent, the white thread underbody on this pattern is important; a dark thread underbody will mute the body color. This fly should be fished drag-free close to the bottom. In shallow water, it can be used as a dropper behind a dry fly that is sufficiently buoyant to support the weighted hook. In deeper water, fish it below an indicator, adding weight to the leader as needed to get the fly near the streambed.

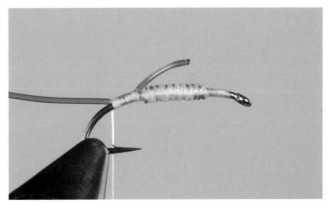

**1.** Weight the middle half of the hook shank using the instructions for the Pheasant Tail Nymph, steps 1–2, p. 21. Secure the wire wraps with the white thread. Mount a strand of the body cord atop the shank behind the rearmost wrap of the wire, using very tight thread wraps. Gently stretch the cord along the top of the shank, and secure it with smooth, even thread wraps, forming a white thread underbody to the rear of the shank. Bind the cord with 3 to 4 tight wraps at the rear of the shank.

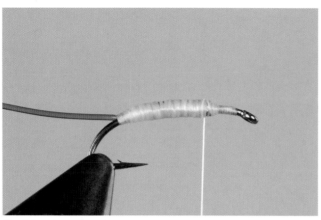

**2.** Trim the excess body cord. With the thread, form a slightly tapered underbody over the rear $^2/_3$ of the shank. Position the thread $^1/_3$ of a shank length behind the hook eye.

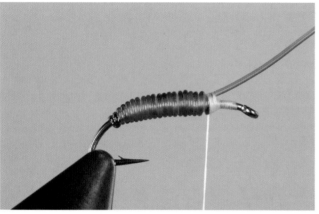

**3.** Gently stretch the cord, and wrap it forward in close, touching turns. Secure the cord at the hanging thread with tight wraps.

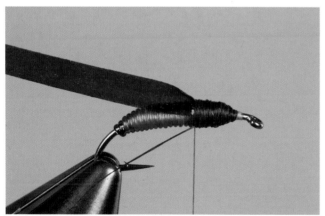

**4.** Clip and bind the excess cord. Tie off and clip the white thread. Mount the brown thread behind the hook eye, and wrap it rearward over the front edge of the abdomen. Mount a strip of Scud Back atop the front edge of the abdomen. Clip and bind the excess. Mount a 4-inch length of ribbing thread over the rearmost mounting wraps securing the Scud Back. Position the thread at the rearmost thread wrap securing the Scud Back.

**5.** Dub a thorax slightly larger in diameter than the front of the abdomen, stopping 4 thread wraps' distance behind the hook eye.

**6.** Fold the Scud Back over the dubbing, and secure it tightly directly ahead of the thorax. Spiral the ribbing thread 2 or 3 times over the thorax; tie off the rib directly ahead of the thorax.

**7.** Clip the excess Scud Back and ribbing thread. Finish the head of the fly. Use a dubbing needle or teaser to pick out some dubbing fibers on the sides of the thorax for legs.

# Spotted Sedge Pupa Patterns

| | |
|---|---|
| **Hook:** | #12-14 scud or 2XL curved nymph |
| **Weight:** | Lead or nontoxic wire |
| **Thread:** | Dark brown 6/0 or 8/0 |
| **Rib:** | Fine copper wire |
| **Shellback:** | Brown or tan ⅛-inch Scud Back |
| **Abdomen:** | Hare's ear or olive-brown Hareline Hare's Ear Ice Dub |
| **Thorax:** | Fox squirrel fur strip |

The ribbed shellback on this pattern gives the fly a fairly realistic segmented appearance. The sparkle dubbing gives it a shimmer in the water like the pupal skin of the natural insect, and the soft fur legs provided by the thorax fur have good movement in the water. This fly can be fished anywhere in the water column—below an indicator in deeper water or as a dropper behind a dry fly to hold it closer to the surface. This fly has some weight to it, however, and a good, buoyant dry fly, such as the Deer Hair Caddis used to imitate the adult, is necessary to support the pupa dropper.

**1.** Beginning at a point about ⅗ of a shank length behind the hook eye, wrap the wire forward to a point ⅕ of a shank length behind the eye, as described for the Pheasant Tail Nymph, steps 1–2, p. 21. Secure the wire with thread. Position the thread at the rear of the shank, and mount a strip of Scud Back and a length of ribbing wire. Clip and bind the excess of both materials. Position the thread at the rear of the shank.

**2.** Dub the abdomen over the rear ⅔ of the hook shank. Position the thread in front of the abdomen. Use a dubbing needle or teaser to pick out some dubbing fibers along the length of the abdomen, and preen these fibers rearward.

**3.** Form and wax a dubbing loop 2 to 3 inches in length, and hang a dubbing hook or twister from the loop as shown in the instructions for the CDC and Wire Caddis, steps 4–5, pp. 136–137. Wrap the tying thread forward, stopping 5 to 6 thread wraps' distance behind the eye.

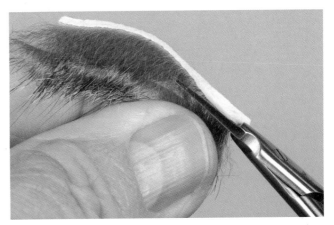

**4.** Cut a pinch of fox-squirrel fur from the hide. After trimming the fur, handle it carefully so that the fibers remain aligned in parallel.

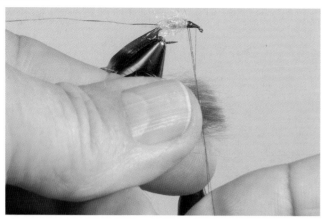

**5.** Insert your right forefinger into the dubbing loop to hold it open. Place the fur in the open loop, perpendicular to the threads, about an inch below the hook shank. Leaving this bare thread will allow unrestricted spinning of the fur.

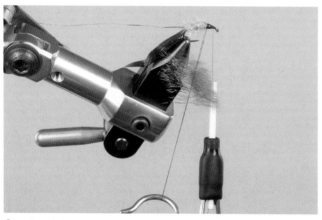

**6.** Carefully remove your finger to close the loop while pulling gently downward on the dubbing hook or twister to keep light tension on the loop threads.

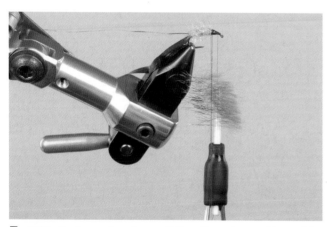

**7.** With the loop closed, carefully adjust the position of the fur by pulling on the tips or butts. The guard hair tips should extend outward from the thread about the length of the abdomen.

**8.** Use the dubbing hook to spin the thread loop in a clockwise direction (as viewed from above), until a spiky, brushlike strand is formed. Do not overspin the loop, or you may break the thread.

**9.** Wrap the dubbing strand forward over the thorax area. As you make each new wrap, use your left fingers to preen back the fibers from the previous wrap. Wrap forward to the hanging thread, and secure the loop.

**10.** Trim the excess dubbing loop, and position the thread at the front of the thorax. Preen the fur on the sides and top of the thorax downward on each side of the shank. Draw the Scud Back over the top of the thorax. Secure it in front of the thorax.

**11.** Clip and bind the excess Scud Back. Spiral the copper wire over the body in 7 to 8 turns. Secure the wire in front of the thorax. Clip and bind the excess, and finish the fly. Use a dubbing needle to pick out any thorax fur trapped beneath the ribbing wire.

## BEADHEAD CADDIS PUPA

| Hook: | #12-16 1XL nymph |
|---|---|
| Head: | Gold or black bead |
| Weight: | Lead or nontoxic wire |
| Thread: | Brown 6/0 or 8/0 |
| Rib: | Medium or fine gold tinsel |
| Abdomen: | Natural Hare-Tron dubbing |
| Wing buds: | Dark brown mottled hen-hackle barbs |
| Thorax: | Dark brown Hare-Tron dubbing |

This simple fly is an excellent producer and a top choice for fishing in fast riffles or pocketwaters when you need a fly that sinks rapidly. In the gold-bead version, it's also quite effective just before and during a sparser Spotted Sedge hatch. Possibly the shiny bead calls more attention to the fly when fewer naturals are in the water. It can be fished beneath a strike indicator or used as a dropper behind a dry fly, though the dry fly must float well, since this pattern is weighted with both wire and a metal bead. Only the forward portion of the shank is weighted in the following sequence, but you can weight the entire shank for an extra-heavy version of the fly.

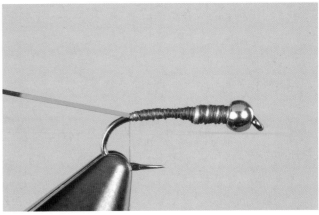

**1.** Weight the hook with wire as shown in the instructions for the Beadhead Prince Nymph, steps 1–3, p. 141. Wrap the thread to the rear of the shank, and mount a length of ribbing tinsel. Position the thread at the rearmost thread wrap.

**2.** Dub the abdomen ¾ of the distance to the rear of the bead. Position the thread ahead of the abdomen.

**3.** Spiral the tinsel over the dubbing in 3 to 4 wraps. Tie off the tinsel ahead of the abdomen.

**4.** Clip the excess tinsel. Align the tips of 8 to 12 hen-feather barbs, and mount them on the far side of the shank, directly ahead of the abdomen, to form wing buds that extend rearward to the end of the body. Mount a second bundle of barbs on the near side of the shank to match the first in length, as shown here from above.

**5.** Clip and bind the feather-barb butts. Dub a thorax to the rear of the bead, and finish the fly behind the bead. Use a dubbing needle or teaser to pick out some fibers from the sides and underside of the thorax to simulate legs. Trim any excessively long fibers.

# Spotted Sedge Emerger Patterns

## PARTRIDGE CADDIS EMERGER

*Originator: Mike Lawson*

| | |
|---|---|
| **Hook:** | #12-16 standard dry-fly |
| **Thread:** | Brown 6/0 or 8/0 |
| **Tail:** | Tan Antron yarn fibers |
| **Rib:** | Fine gold wire |
| **Abdomen:** | Tan dubbing |
| **Underwing:** | Tan Antron yarn fibers |
| **Overwing:** | 2 brown partridge feathers |
| **Thorax/head:** | Peacock herl |

This pattern can be used to imitate other types of caddis by changing the body color and hook size to match the natural, but the partridge-feather wings are a particularly good match for the mottled wings of the Spotted Sedge. This fly is designed to be fished subsurface, by itself or as a dropper behind a dry fly, and this slow-sinking pattern is well suited for slow-moving, glassy water. From the standpoint of representation, the peacock-herl head is a little puzzling, but there's no denying the effectiveness of the pattern.

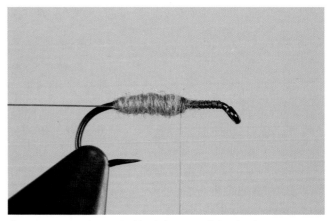

**1.** Mount the thread behind the eye of the hook, and wrap a foundation to the rear of the shank. Mount a length of gold ribbing wire. Then dub a slightly tapered abdomen over the rear ²/₃ of the hook shank.

**2.** Cut a 2- to 3-inch length of Antron yarn. Mount the yarn at the middle of the strand atop the shank directly in front of the abdomen. Do not clip the excess. Draw the rear part of the strand back over the top of the abdomen, and hold it in place with your left hand. With your right hand, secure the yarn atop the shank, just behind the abdomen, with 2 firm turns of the ribbing wire.

**3.** Spiral the ribbing wire forward, securing the yarn strand to the top of the abdomen as you wrap.

**4.** Continue wrapping the ribbing forward until you reach the front of the abdomen. Tie off the ribbing atop the thread wraps used to mount the yarn strand. Clip the excess ribbing wire. Fold the tag end of the yarn rearward, over the top of the hook. Take several tight thread wraps over the fold, wrapping rearward, to secure the underwing. It should slant low over the back of the fly.

**5.** Use thread wraps to even out any inconsistencies in the mounting wraps used to secure the fold of the yarn, forming a smooth foundation for mounting the wings. Position the thread at the front of the abdomen. Clip the underwing so that it extends just to the end of the abdomen, and trim the shuck fibers to the length of the abdomen. Select 2 matched partridge feathers. Preen the lower barbs toward the butts of the feathers, so that the tip portions are as long as the hook shank.

**6.** Stack the feathers with the tips aligned and the curvatures opposed, so that the feathers flare away from one another. Mount the feathers edgewise (not flat) at the front of the abdomen so that they lie side by side atop the shank, as shown in this top view.

**7.** Clip and bind the feather butts. Position the thread at the rearmost wing-mounting wrap. Mount 2 or 3 strands of peacock herl and a 6-inch length of scrap tying thread. Clip and bind the excess. Position the thread 3 thread wraps' distance behind the hook eye.

**8.** Twist the herl and thread together, and wrap the thorax of the fly as described for the Pheasant Tail Nymph, steps 8–9, pp. 22–23. Wrap forward to the hanging thread. Secure the herl strand, and clip the excess. Finish the fly.

## X-CADDIS

*Originator: Craig Mathews*

| | |
|---|---|
| **Hook:** | #12-16 standard dry-fly |
| **Thread:** | Brown 6/0 |
| **Tail:** | Tan Antron yarn fibers |
| **Body:** | Tan or olive-brown dubbing |
| **Wing:** | Deer hair |

This fly represents a fully emerged adult with the pupal shuck still attached at the rear. By eliminating the palmered hackle, this version of the Elk Hair Caddis strikes a useful compromise. The body and the tail, which suggest a trailing shuck, sit tight against the film, and the wing presents a clean, unobstructed profile, making this fly highly effective on flat water. At the same time, the deer-hair wing provides plenty of flotation, and it's buoyant enough to fish in turbulent water or to support a weighted pupa pattern. That it's quick and easy to tie makes this pattern even better.

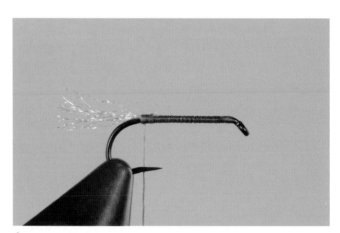

**1.** Attach the thread behind the hook eye, and wrap a foundation to the tailing point. Mount a length of Antron yarn atop the shank; clip and bind the excess. Trim the tail to one hook gap in length.

**2.** Dub a slightly tapered body over the rear ²/₃ of the shank, and position the thread directly in front of the body.

**3.** Clean and stack a bundle of deer hair as described for the Sparkle Dun, steps 1–4, pp. 27–28. Size the hair atop the hook shank so that the hair tips extend to the end of the body.

**4.** Mount the wing atop the shank, directly ahead of the body, as described for the Compara-emerger, steps 3–6, p. 40.

**5.** Finish the fly around the hook shank beneath the hair butts. Trim the hair butts following the angle of the hook eye to leave a tuftlike head.

## Spotted Sedge Adult Patterns

**DEER HAIR CADDIS**

| | |
|---|---|
| **Hook:** | #12-16 standard dry-fly |
| **Thread:** | Brown 6/0 |
| **Body hackle:** | Dun dry-fly |
| **Body:** | Fine tan or olive-brown dubbing |
| **Wing:** | Dark deer hair |

This workhorse pattern is a staple in our fly boxes. For a bushy fly, it works surprisingly well on flat water, but it's really at its best where a more buoyant pattern is needed—in rough, broken, or choppy water, or as a point fly to support a weighted or bead-head dropper. The color and mottled appearance of the deer-hair wing are a good match for the Spotted Sedge. The instructions for tying this fly are shown on pp. 131–133.

## EZ CADDIS

Originator: Mike Lawson

| | |
|---|---|
| **Hook:** | #12-16 standard dry-fly |
| **Thread:** | Olive 8/0 |
| **Wing post:** | Calf body hair |
| **Body:** | Fine tan or olive-brown dubbing |
| **Wing:** | Brown partridge feather |
| **Hackle:** | Dyed-dun grizzly or natural grizzly dry-fly |

This flush-floating design works equally well on calm or choppy water for selective fish. It has enough flotation to support an unweighted pupa or emerger pattern as a dropper. A big advantage of this fly is the white wing post, which makes it easy to see in the low light of evening, when the Spotted Sedges commonly hatch. The original dressing calls for a pair of partridge feathers to form the wing; we use a single feather with the tip removed to make the EZ Caddis even easier.

**1.** Form the wing post as described for the Parachute Adams, steps 1–2, p. 97.

**2.** Prepare and mount the hackle feather as shown in the instructions for the Parachute BWO, steps 3–4, p. 16. Position the thread at the rear of the hook.

**3.** Dub the abdomen to the base of the wing post, then wrap the thread tightly rearward, back over the dubbing, to the midpoint of the shank. Position the thread at the midpoint of the shank. Building up an underbody of dubbing beneath the wing-mounting point will help ensure that the wing feather lies low across the back of the fly.

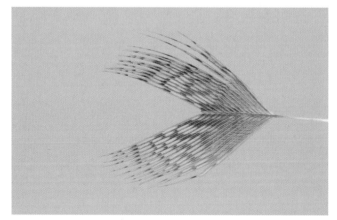

**4.** Select a partridge feather, and snip the stem near the tip of the feather to leave a V-shaped gap. The uppermost feather barbs should be about as long as the hook shank. Then strip away the lower barbs to leave a segment of barbs about 1 to 1½ hook gaps in width on each side of the stem.

**5.** Mount the feather, concave side down, atop the shank to form a wing one shank length long.

**6.** Bind and clip the wing butts. With a dubbed thread, cover the wing-mounting wraps and form the thorax, stopping 4 thread wraps' distance behind the hook eye.

**7.** Wrap and tie off the hackle as described for the Parachute BWO, steps 7–10, pp. 16–17. Finish the fly.

# Spotted Sedge Egg-Laying Caddis Patterns

## CANOE FLY

| | |
|---|---|
| **Hook:** | #12-16 standard dry-fly |
| **Thread:** | Brown 6/0 |
| **Body:** | Olive-brown dubbing |
| **Wing:** | Deer hair |

This simple, two-material fly can be fished dead-drift on the surface to imitate a flush-floating, spent caddis. Or it can be fished to imitate the diving female: let the fly float drag-free, and at the end of the drift, give the rod tip a sharp twitch to pull the fly underwater. Then use a pulsing motion of the rod tip to swim the fly down and across the current.

**1.** Mount the thread behind the hook eye, and wrap a foundation to the rear of the shank. Dub a body over the rear $2/3$ of the shank, and position the thread in front of the body.

**2.** Clean and stack a bundle of deer hair as described for the Sparkle Dun, steps 1–4, pp. 27–28. Position it atop the hook shank so that the hair tips extend just beyond the hook bend.

**3.** Mount the wing atop the shank, directly ahead of the body, as described for the Compara-emerger, steps 3–6, p. 40. Finish the fly around the hook shank beneath the hair butts. Trim the hair butts at the same angle as the hook eye.

## HARE'S EAR FLYMPH

*Originator: Jim Leisenring*

| | |
|---|---|
| **Hook:** | #12-16 standard nymph |
| **Thread:** | Red Pearsall's Gossamer Silk or 6/0 |
| **Tail:** | Mottled brown hen-hackle barbs |
| **Rib:** | Oval gold tinsel |
| **Body:** | Hare's mask dubbing |
| **Hackle:** | Mottled brown hen |

This more heavily dressed version of the soft-hackle design is usually fished wet-fly style—casting down and across and letting the fly swing across the current—to represent an emerging insect or swimming adult caddis. But an equally effective, if less commonly practiced, technique is to present the fly dead-drift to imitate a sunken, spent female. Fishing it as a dropper behind a dry fly or small yarn indicator greatly improves strike detection in the waning light, when Spotted Sedges are most often on the water.

**1.** Mount the thread behind the hook eye, and position it 3 thread wraps' distance behind the eye. Strip the short, fuzzy barbs from the base of the hackle feather. Mount the feather edgewise atop the shank, as shown, with the front or convex side of the feather facing you. Bind the stem atop the shank with 4 to 5 tight wraps moving rearward. Clip the excess, and wrap the thread to the tailing point.

**2.** Align the tips of 6 to 8 hen-hackle barbs, and mount them atop the shank to form a tail one shank length long. Mount a length of ribbing tinsel over the rearmost tail-mounting wraps. Clip and bind the excess of both materials.

**3.** Position the thread about 4 thread wraps' distance ahead of the rearmost tail-mounting wrap. Form and wax a dubbing loop as shown in the instructions for the CDC and Wire Caddis, steps 4–6, pp. 136–137. Position the thread ⅓ of a shank length behind the hook eye. Hold the loop open with your left index finger; hold a clump of dubbing between your left thumb and middle finger. With your right hand, apply small pinches of dubbing to one of the loop threads. The wax will hold the dubbing in position. Begin applying dubbing near the hook shank, placing each pinch below the previous one until you've covered 2 to 3 inches of the loop thread.

**4.** Spin the dubbing hook or twister clockwise (as seen from above) to form a rough, spiky cord. Wrap the dubbing forward over the rear ⅔ of the shank, and secure the strand with the tying thread.

**5.** Trim off the excess dubbing strand, and position the thread at the front of the body. Spiral the tinsel forward in 3 to 5 turns to the front of the body. Tie off and clip the excess tinsel. Then wrap the tying thread rearward 1 to 2 turns, to a point just forward of the middle of the body.

**6.** Take 2 turns of hackle, wrapping rearward.

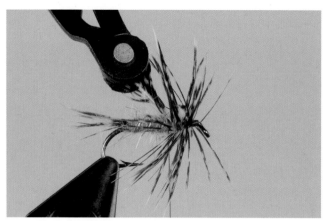

**7.** Take a third wrap of hackle rearward, halfway to the point where the tying thread is positioned. Take a fourth wrap of hackle directly in front of the tying thread. Secure the hackle tip directly over the body with 2 tight wraps of thread.

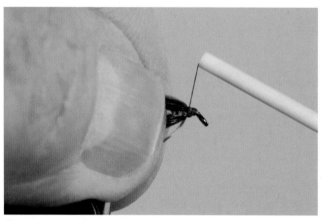

**8.** Spiral the thread forward through the wrapped hackle, wiggling the thread back and forth as you wrap to avoid binding down any of the hackle barbs. Position the thread in front of the first hackle wrap. Clip the excess hackle tip. Use your left fingers to smooth the hackle rearward, and form the head of the fly against the base of the hackle to slant the fibers rearward.

# LOCALLY IMPORTANT HATCHES

## Fall Caddis

These caddisflies are found in a fair number of streams in the Rocky Mountain region but are most common in the Pacific Coast states, where they inhabit waters with moderate to fast-flowing currents. The larvae construct cases from sand and small pebbles, and the adults generally hatch in the autumn months, hence the name. The presence of these caddis around a trout stream is almost impossible to overlook. They are big, showy insects, often the largest insect of the season on streams where they appear, and trout fattening up for the winter months don't fail to take advantage of them. Populations can vary significantly from stream to stream, but even sparser hatches bring up large fish. Skating an adult imitation on a Pacific Coast stream has surprised more than one angler with a steelhead taking the fly.

**Other common names:** Giant Orange Sedge, October Caddis
**Family:** Limnephilidae
**Genus:** *Dicosmoecus*
**Emergence:** September to November, late afternoon to after dark
**Egg-laying flight:** Afternoon until dark
**Body length:** ¾–1 inch (20–25 mm)
**Hook sizes:** #6-8

**Above:** *Fall Caddis larva: slightly curved case formed of sand or small pebbles; creamy yellow to brown body.*

**Above:** *Fall Caddis pupa: orange-brown to pale yellow body; dark brown wing pads.*

**Left:** *Fall Caddis adult: brown wings with gray tones.*

**Above:** *Fall Caddis adult (underside): orange with reddish brown tint.*

# Important Fishing Stages

**Larva** Even though trout will take these larvae, the insects are not found reliably enough in the drift to constitute a fishable phase of the life cycle.

**Pupa** At maturity, the larvae migrate to quiet waters along the shoreline, where they attach to rocks and pupate. At emergence time, which usually occurs from late evening into the night, the pupae swim or crawl to exposed rocks. They cling to rocks at or just above the waterline, and the adults emerge. In this process, some pupae are swept away by the current, and though they normally swim back to shore to emerge, some pupae do end up hatching in open water. The hatch cycle generally lasts for two to three weeks, and a pupa of this size caught in the drift does not go unnoticed. Fishing a pupa imitation at any depth from late afternoon until dark during the hatch season is an effective approach.

**Emerger/adult** Because most fall caddis emerge at the shoreline or on exposed rocks, the emergers and newly hatched adults are not ordinarily found on the water or available to trout.

**Egg-laying female** Although you may occasionally find an adult emerging in open water, the fall caddis you see fluttering on or near the water are normally females returning to lay their eggs. The egg-laden females drop onto the surface of the water for a few moments, wiggle their abdomens to release some of the eggs, then fly off the water to a new location, where they repeat the process. This flighty behavior causes trout to take them aggressively, as soon as females hit the water. As a rule, you don't see a large number of rises during this hatch, and most often you'll be casting to likely holding water rather than visibly feeding fish. But the flies are large enough that it doesn't take many of them to get the attention of the trout. To imitate these ovipositing females, twitch the fly occasionally during a drag-free float, and at the end of the drift, skate the fly across the surface before picking it up to make another cast.

# Fall Caddis Pupa Pattern

**BIG ORANGE PUPA**

| | |
|---|---|
| **Hook:** | #6-8 2XL nymph |
| **Weight:** | Lead or nontoxic wire |
| **Thread:** | Brown 6/0 |
| **Body:** | Orange-brown Antron dubbing |
| **Wing pads:** | Dark brown wing-quill sections, coated with Dave's Flexament |
| **Head:** | Dark brown coarse dubbing |

The orange body and contrasting dark wing pads make the Fall Caddis Pupa quite distinctive, and these features are reasonably easy to replicate on the hook shank. Don't neglect to pick out the body dubbing in step 2, which creates a sparkly halo around the abdomen like the pupal skin of the natural. Similarly, it's important to pick out the head dubbing in step 4, since the pupa has long legs held close to the body. You can fish this fly at any depth, beneath an indicator or as a dropper behind a dry fly, but concentrate your efforts in water directly downcurrent from exposed rocks in the stream and swifter currents adjacent to calmer, rocky banksides—areas where the pupae are most apt to be dislodged by the current.

**1.** Use wire to weight the hook, as shown in the instructions for the Pheasant Tail Nymph, steps 1–2, p. 21. For this pattern, however, wrap the wire over the middle half of the shank. Position the thread at the rear of the hook shank.

**2.** Dub the body over the rear ¾ of the shank. When finished, take 3 to 4 tight thread wraps rearward over the dubbing at the very front of the body. Mounting the wings over this dubbed underbody will help ensure that they hug the sides of the fly rather than flare out. Use a dubbing needle or teaser to pick out some dubbing fibers around the body. Trim any excessively long fibers.

**3.** Mount the wing pads as described for the Latex Caddis Pupa, steps 6–9, p. 125. Position the thread atop the rearmost wrap securing the wing pads.

**4.** Dub a head to just behind the hook eye, and finish the fly. Use a dubbing needle or teaser to pick out longer fibers on the underside of the shank for legs.

# Fall Caddis Adult Pattern

## DARK CADDIS

*Originator: Polly Rosborough*

| | |
|---|---|
| **Hook:** | #6-8 standard dry-fly |
| **Thread:** | Black 6/0 |
| **Body hackle:** | Dark brown dry-fly |
| **Body:** | Orange dubbing |
| **Wing:** | Dark brown or natural dark deer hair |
| **Hackle:** | Dark brown dry-fly |

The colors of this fly match the orange and amber coloration of the natural insect, but it's really the design that makes this pattern effective. The deer hair creates the hazy, indistinct silhouette of fluttering wings, and the abundance of hackle makes this fly sit lightly on the water so that it can be twitched and skated. The wraps of hackle palmered over the body are placed close together; if possible, use a saddle hackle to make sure the feather has enough length for these wraps, particularly on the larger hook size. This pattern floats well and can be fished in any type of water. Dead-drift it with occasional twitches, or skate it across the surface with pauses. Keep the fly treated with floatant so that it sits up high on the water; a low-riding or waterlogged fly will plow through the water rather than skimming over the surface when skated.

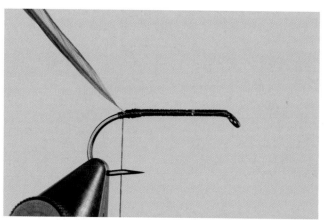

**1.** Mount the thread behind the hook eye, and wrap a thread foundation, stopping about 3 thread wraps' distance ahead of the normal tailing point on the hook. Prepare and mount a hackle as described for the Little Green Drake, steps 4–6, p. 42. When mounting the feather, leave a length of bare hackle stem between the rearmost mounting wrap and the lowermost barbs on the feather. The length of this bare stem should be equal to ½ the thickness of the dubbed abdomen. Position the thread just behind the hackle.

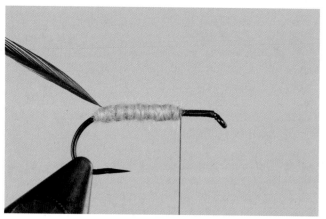

**2.** Beginning behind the mounted hackle, dub the abdomen over the rear ²/₃ of the shank. Position the thread directly ahead of the abdomen.

**3.** Take 8 to 10 wraps of hackle over the body, and tie off the hackle, as described for the Deer Hair Caddis, steps 3–5, p. 132. Clip the hackle tip, and position the thread directly in front of the body.

**4.** Clean and stack a bundle of deer hair as shown in the instructions for the Sparkle Dun, steps 1–4, pp. 27–28. Mount the hair atop the shank to form a wing that extends rearward to the hook bend. Secure the hair tightly.

**5.** Clip the butts of the wing material. Flatten the thread by spinning the bobbin counterclockwise (as viewed from above). Bind down the deer-hair butts, creating a thread foundation that tapers evenly toward the hook eye. Then prepare and mount a hackle over the rearmost thread wrap securing the wing as shown in the instructions for the Little Green Drake, steps 4–6, p. 42. Position the thread about 4 thread wraps' distance behind the hook eye.

**6.** Wrap the hackle forward to the hanging thread, and tie off the feather as described for the Little Green Drake, steps 7–10, pp. 42–43. Clip the excess and finish the fly.

# Chapter Three

# *Stoneflies*

## MAJOR HATCHES

## Golden Stonefly

Golden Stonefly nymphs inhabit the rocky substrate in riffles and runs that have moderate to fast current flows. These are large bugs, and the nymphal stage can last two to three years, depending on the water temperature and availability of food. Moreover, these nymphs are predators that crawl about the bottom looking for smaller insects on which to feed, and in clambering about, they expose themselves to the force of the current more often than less active aquatic organisms. This predaceous behavior, combined with the long maturation period, means that Golden Stonefly nymphs can be found in the drift at any time of the year, particularly on streams with large populations. Mature nymphs crawl out of the water onto land, where the adults hatch, so the actual emergence provides little in the way of dry-fly opportunity. But after mating, the females return to the water to deposit their eggs, skimming across or resting on the surface, and trout are quick to take advantage of this opportunity. Golden Stoneflies emerge en masse; a hatch can lasts for 2 to 4 weeks, and even longer in some waters, and typically begins in the lower sections of streams, moving upriver as the water warms. Perhaps because of their larger size, Salmonflies tend to steal the limelight from their smaller cousins, but many anglers consider the Golden Stoneflies to be the best stonefly hatch in the West: the active nymphs provide year-round subsurface fishing, and the adults produce some of the season's best dry-fly fishing for big trout.

| | |
|---|---|
| **Other common name:** | Golden Stone |
| **Family:** | Perlidae |
| **Genera:** | *Hesperoperla* and *Calineuria* |
| **Emergence:** | May to August, after dark |
| **Egg-laying flight:** | Midday to early evening |
| **Body length:** | 1–1½ inches (25–38 mm) |
| **Hook sizes:** | #6-8, 2XL–3XL |

**Left:** *Golden Stonefly nymph: brown to yellow-tan with distinct mottling.*

**Above:** *Golden Stonefly nymph (underside): yellow-tan to light brown.*

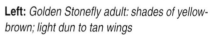

**Left:** *Golden Stonefly adult: shades of yellow-brown; light dun to tan wings*

**Above:** *Golden Stonefly adult (underside): gold to yellow-brown.*

# Important Fishing Stages

**Nymph** These restless nymphs are available to trout throughout the year, but it's during the emergence, when the nymphs are migrating toward shore, that the largest number are swept off the bottom and into the current. Built for crawling, with strong, stout legs, Golden Stonefly nymphs are poor swimmers, and if they lose their footing, they're often carried good distances downcurrent before settling to the streambed. Even so, they are likely to drift closer to the bottom rather than higher up in the water column, and a weighted pattern fished deep in riffles and runs is generally the most productive approach. During an emergence period, the migrating nymphs congregate in waters near the bank and wait for the cover of darkness before crawling out onto dry land. At these times, pay special attention to current bands and seams close to the shore.

**Adult** After dark, the mature nymphs crawl from the water and climb onto adjacent rocks or vegetation, where the adults hatch. Mating takes place in streamside foliage. A day or two after emerging, the females return to the water from midday to early evening and deposit their eggs by repeatedly dipping to the surface. Afterward, most of the females fall spent to the water. For both trout and angler, these egg-laying flights are a golden opportunity; since ovipositing can take place near the banks, in the middle of the river, or on the water in between, fishing a dry fly to a likely looking spot anywhere in the river may bring the kind of explosive strike that is one of the thrills of the Golden Stonefly hatch. When you observe no insects on or above the water, focus your efforts near the bank, beneath overhanging vegetation. Several males may cluster around a female in competition for mating, and in the scramble, they often end up in the water. Breezy days can be particularly good. Trout have a long memory of these large insects and will frequently take an adult pattern for a week or two after the hatch has ended. When prospecting, using a nymph dropper behind the dry fly is very effective, especially in heavily fished waters.

# Golden Stonefly Nymph Patterns

## BROWN SHELLBACK STONEFLY

| | |
|---|---|
| **Hook:** | #6-8 2XL standard nymph |
| **Weight:** | Lead or nontoxic wire |
| **Thread:** | Brown, 6/0 |
| **Tails:** | Brown rubber leg material |
| **Shellback/wing case:** | Mottled Oak Natural Thin Skin |
| **Rib:** | Tan vinyl ribbing |
| **Abdomen:** | Medium brown, light brown, or golden brown dubbing |
| **Thorax:** | Medium brown, light brown, or golden brown dubbing |
| **Legs:** | Brown rubber leg material |

A freshly molted Golden Stonefly nymph can temporarily exhibit some striking yellow or gold markings over its entire body. But these bright colors don't last long, and most nymphs are a more subdued shade on top, such as the brown on this pattern. The shellback and wing-case material gives a mottling to the dorsal side of the fly that is characteristic of the natural, and you can choose a dubbing color to match the undersides of the naturals in the waters you fish. This shell-back style of nymph has been around for quite some time; years ago, tiers formed the shellback from a lacquered hen-saddle or gamebird feather—an approach that was both time-consuming and messy. When synthetic films such as Thin Skin became available, tiers switched to them immediately, with no decrease in the effectiveness of the finished fly but a significant increase in the speed and simplicity of dressing it. Thin Skin ties quite easily, but when tapering the shellback strip as shown in step 2, leave the material attached to the paper backing; it cuts more easily and accurately. The amount of wire used to weight the fly can be varied to make a heavier or lighter pattern than is shown here.

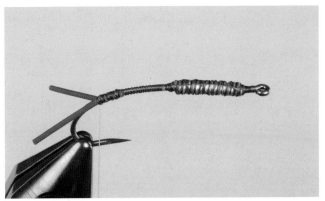

**1.** Wrap and secure a wire underbody over the front half of the hook shank as described for the Pheasant Tail Nymph, steps 1–2, p. 21. Wrap a thread foundation to the rear of the shank. Mount a length of rubber leg material on each side of the shank to form split tails. Clip and bind the excess tail material, and then trim the tails to be one hook gap in length. Position the thread at the rear of the shank.

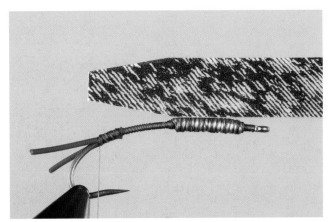

**2.** Cut a strip of Thin Skin 2 inches long and one hook gap in width. Beginning at a point ½ a shank length from the end of the strip, taper the strip on both sides so that the end of the strip is ½ the hook gap in width.

**3.** Mount the Thin Skin strip, glossy side up, by the narrow end directly atop the rearmost thread wraps securing the tail. Then mount a length of ribbing material atop the rearmost thread wraps securing the Thin Skin. Clip and bind the excess ribbing. Dub a slightly tapered abdomen to the midpoint of the shank. Position the thread at the front of the abdomen.

**4.** Draw the strip of Thin Skin forward smoothly over the top of the abdomen. Secure the strip with 3 tight thread wraps directly in front of the abdomen, as shown in this top view, but do not clip the excess.

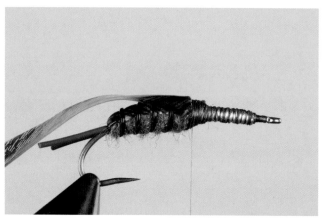

**5.** Spiral the ribbing material over the abdomen in 4 to 5 evenly spaced wraps. Secure the ribbing in front of abdomen. Clip and bind the excess ribbing. Fold the Thin Skin strip rearward, and wrap the thread rearward over the fold until the thread abuts the front of the abdomen.

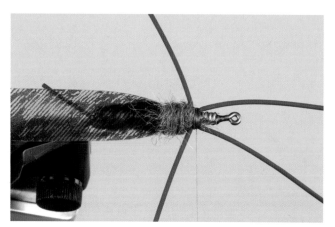

**6.** Dub a thorax slightly larger in diameter than the abdomen halfway to the hook eye. Mount a strand of rubber leg material on each side of the shank as described for the Green Drake Rubber Legs, steps 6–9, pp. 58–59, and shown in this top view.

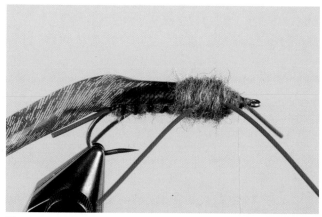

**7.** Dub over the leg-mounting wraps, and then dub the remainder of the thorax forward, ending about 5 thread wraps' distance behind the hook eye. Position the thread directly in front of the thorax.

**8.** Fold the Thin Skin forward smoothly over the top of the thorax. Secure it tightly in front of the thorax. Clip and bind the excess, and finish the head of the fly. Trim the legs to about 2/3 the length of the hook shank.

## BASIC BROWN STONE

| | |
|---|---|
| **Hook:** | #6-8 2XL–3XL nymph |
| **Weight:** | Lead or nontoxic wire |
| **Thread:** | Dark brown 6/0 |
| **Tails:** | Brown rubber leg material |
| **Rib:** | Copper wire |
| **Body:** | Brown dubbing |
| **Wing case:** | Turkey feather section, coated with Dave's Flexament, or dark brown Antron yarn |
| **Legs:** | Brown rubber leg material |

In our experience, highly realistic, anatomically correct stonefly nymph representations do not pay a reward in trout that justifies the time and effort spent in dressing the patterns. A suggestively buggy pattern such as the Basic Brown Stone often gets better results, not so much because of the appearance of the pattern, but, strange as it may sound, because the fly can be lost to a snag without much misgiving. It may seem odd to talk about expendability as a merit in a trout fly, but when fishing stonefly nymphs, it's important to replicate the drift of the natural insects—that is, drag-free and close to the bottom. The fly should touch the streambed a few times during the drift, and when fished like this, it will occasionally hang up, irretrievably, on the bottom. No one likes to lose flies, especially those that are time-consuming or laborious to tie, so anglers are sometimes reluctant to fish a stonefly nymph as close to the bottom as they should for maximum effectiveness. Because this simple pattern is relatively quick to dress and made of inexpensive materials, you can fish it deep even in snaggy waters without hesitation. Get this fly where it belongs, and it catches trout quite well. The lacquered turkey feather provides a glossy, shell-like wing case, but the fly can be dressed instead with a dark brown Antron yarn wing case, which works just as well and is even quicker and easier to tie.

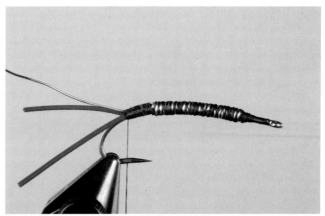

**1.** Wrap an underbody of wire over the middle $^2/_3$ of the hook shank, and form a tapered ramp of thread at each end of the wire underbody as described for the Pheasant Tail Nymph, steps 1–2. p. 21. Secure the wire with thread wraps, and position the thread at the rear of the shank. Mount a length of rubber leg material on each side of the shank to form split tails about one hook gap in length. Then mount a strand of ribbing wire atop the rearmost thread wrap securing the legs. Position the thread at the rearmost wrap securing the tails and ribbing.

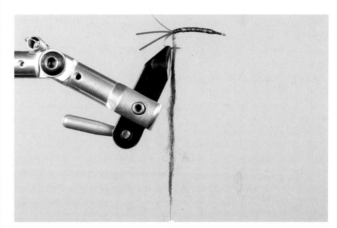

**2.** Wax about 4 to 6 inches of the tying thread, and apply dubbing to the thread in an even layer. Don't dub the thread too heavily; use a moderate amount of material. Leave about 1 inch of bare thread between the base of the dubbing and the tip of the bobbin.

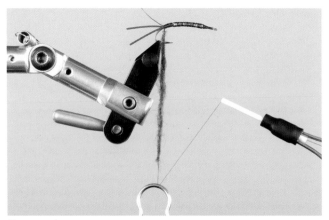

**3.** With your left hand, use a dubbing hook to capture the thread at the tip of the bobbin. Put moderate tension on the dubbing hook. Strip off enough thread from the bobbin to reach the hook shank.

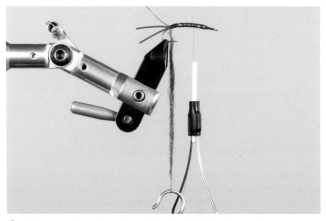

**4.** Return the thread to the hook shank at the rearmost mounting wrap securing the legs and ribbing. Take 3 to 4 tight wraps of thread around the shank, then position the thread at the midpoint of the shank. You've now formed a loop of thread around the dubbing hook, with one side of the loop dubbed and the other bare thread.

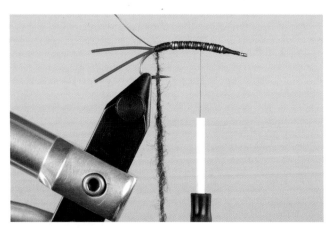

**5.** Spin the handle of the dubbing hook (clockwise as seen from above) to twist the loop strands together, securing and compacting the dubbing.

**6.** Wrap the dubbing loop forward, forming a tapered abdomen. Build the taper in layers. When the entire dubbing loop is wrapped, repeat steps 2–5 to form a second loop, and continue forming the abdomen to the midpoint of the shank. When the abdomen is complete, tie off the dubbing loop.

**7.** Clip the excess dubbing loop. Spiral the ribbing wire forward to the front of the abdomen, wrapping in the opposite direction from the one in which you wrapped the dubbing. Tie off and clip the wire. Cut a section of turkey feather about one hook gap in width, and mount it atop the shank, dull side upward, directly ahead of the abdomen. Wrap rearward to secure the feather atop the front edge of the abdomen. Trim the excess feather.

**8.** Dub a thorax slightly larger in diameter than the abdomen, stopping 6 to 8 thread wraps' distance behind the hook eye. Wrap the thread back to the middle of the thorax. Mount a strand of rubber leg material on each side of the shank as described for the Green Drake Rubber Legs, steps 6–9, pp. 58–59. Once the legs are secure, cover the mounting wraps with a thinly dubbed thread. Position the thread at the front of the thorax.

**9.** Draw the turkey feather forward over the top of the thorax, and secure it in front of the thorax. Trim the excess, and finish the head of the fly. Trim the tails to one hook gap in length and the legs to ½ the shank length.

# Golden Stonefly Adult Patterns

**YELLOW STIMULATOR**

*Originator: Randall Kaufmann*

| | |
|---|---|
| **Hook:** | #6-8 TMC 200R |
| **Thread:** | Orange 6/0 |
| **Tail:** | Elk hair |
| **Rib:** | Fine gold wire |
| **Abdomen hackle:** | Brown dry-fly saddle |
| **Abdomen:** | Yellow Antron dubbing |
| **Wing:** | Elk hair |
| **Thorax hackle:** | Grizzly dry-fly |
| **Thorax:** | Amber Antron dubbing |

This fly is really a version of an older pattern, the Bucktail Caddis, but the palmered hackle and elk-hair wing and tail make it float much better. In fact, it's an ideal fly to use with a weighted or bead-head nymph dropper. Adult Golden Stoneflies, particularly those that end up in the water accidentally, often "run" across the surface film back to the bank rather than becoming airborne and flying to safety. The abundance of hackle on this pattern makes it ride lightly on the water, and under tension, the fly will skate across the surface. An effective method for fishing this fly is to interrupt a drag-free float by twitching the rod tip to make the fly skitter on the water, and then resume the dead drift. This is a versatile pattern that can be altered in size and color to match other stonefly species, larger caddis, and even grasshoppers. Although a bit time-consuming to dress, it's durable and a good producer.

**1.** Mount the thread behind the hook eye, and wrap a tight thread foundation to the rear of the shank. Clean and stack a bundle of elk hair about ⅓ the hook gap in thickness as described for the Sparkle Dun, steps 1–4, pp. 27–28. Mount the bundle atop the shank to form a tail one hook gap in length. Bind the hair butts atop the rear ⅔ of the hook shank. Clip the excess, and return the thread to the tail-mounting point.

**2.** Mount a length of ribbing atop the rearmost tail-mounting wrap, and clip the excess. Dub a slightly tapered abdomen over the rear ⅔ of the shank. Position the thread at the front of the abdomen.

**3.** Select a brown hackle feather with barbs about one hook gap in length. Prepare the feather and mount it directly ahead of the abdomen as described for the Little Green Drake, steps 4–6, p. 42.

**4.** Spiral the feather rearward to the end of the abdomen in evenly spaced wraps. With your left fingers, hold the tip of the feather above the shank and slanted slightly rearward. With your right fingers, take a complete wrap of the ribbing wire around the feather tip, ending with the wire held above the shank and angled slightly forward.

**5.** Spiral the ribbing wire forward, wiggling it slightly from side to side as you wrap in order to prevent the rib from binding down any hackle barbs.

**6.** When you reach the front of the abdomen, secure the wire and clip the excess. Then trim off the hackle tip at the rear of the fly. Clean and stack a bundle of elk hair about one hook gap in thickness. Position the bundle atop the shank so that the tips are aligned with the hook bend.

**7.** Mount the elk hair atop the shank with tight thread wraps directly in front of the abdomen. Trim the hair butts at an angle tapering toward the hook eye, and bind them down with thread. Position the thread at the rearmost thread wrap securing the wing.

**8.** Select a grizzly hackle feather with barbs the same length as those on the brown feather. Prepare the feather and mount it directly over the wing-mounting wraps. After securing and clipping the hackle stem, dub a tapered thorax to within 4 thread wraps' distance of the hook eye. Position the thread at the front of the thorax.

**9.** Spiral the hackle forward in 4 to 5 wraps to the front of the thorax. Tie off the hackle as described for the Little Green Drake, steps 9–10, pp. 42–43. Finish the head of the fly.

## DIRTY-YELLOW CHUGGER

**Hook:**     #6-8 Mustad 94840
**Thread:**   Dark brown 6/0
**Body:**     Yellow Larva Lace Dry Fly Foam
**Cement:**   Dave's Flexament, or similar flexible cement, thinned 1-to-1 with solvent
**Wing:**     Tan over black Poly-Bear Fiber by Spirit River
**Legs:**     Brown rubber material

This pattern has become our favorite imitation for both adult stoneflies and grasshoppers. The segmented foam abdomen is quite light but creates a substantial, blocky profile on the water, and the body rests flush against the surface film like that of the natural insect. At the same time, the wing is elevated enough to make the fly quite visible to the angler, especially if it is topped, as shown in the following sequence, with tan yarn, which lights up on the water. The pattern floats extremely well, and even a sodden fly can be dried with a single sharp false cast that flushes water from the wing. Because part of the foam body is affixed to the underside of the shank, the effective hook gap is narrowed; to ensure reliable hookups, use a wide-gap hook or a slightly larger hook size. We use a flexible cement in many of the tying steps to make the fly more durable; the body may seem loose as it's being tied, but it will become firmly fixed once the glue dries. Instead of the Larva Lace, you can substitute a strip of a similar 1/16-inch (2 mm) closed-cell foam. For information on types of foam suitable for this pattern, see p. 79.

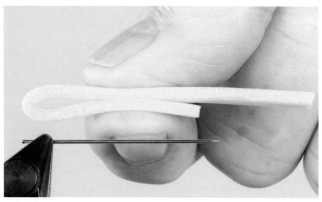

**1.** Mount a sewing needle in the vise. Take a 3½-inch-long strip of ¼-inch-wide closed-cell foam, and fold the strip crosswise so that one side extends 1 inch beyond the other, as shown.

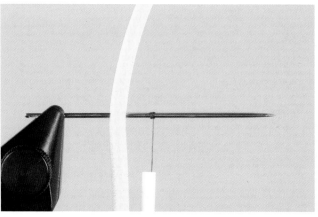

**2.** Position the center of the fold in front of the needle, and then push it onto the needle. Mount the tying thread one hook gap's distance in front of the foam.

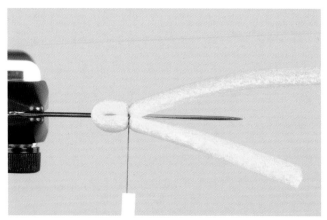

**3.** Fold the foam tags forward, and take a loose thread wrap around the foam; snug it down, and take 4 additional tight wraps. Use 3 half hitches to secure the thread, as shown in this top view.

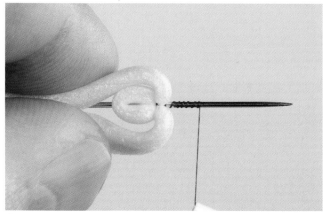

**4.** Fold the foam tags rearward, as shown here from the top, and spiral the thread forward a distance of one hook gap.

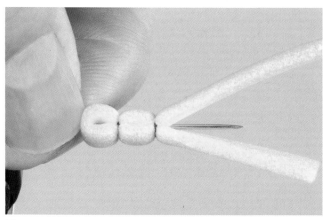

**5.** Fold the foam tags forward, and take a loose thread wrap around the foam; snug it down, and take 4 additional tight wraps. Take 3 half hitches, and trim the thread. Use your thumb and forefinger to push the body extension off the needle.

**6.** On the inside face of the longer foam tag, measure one hook gap's distance from the end of the second segment formed. Position the hook point at the center of the strip, and push it down through the foam. Mount the hook in the vise.

**7.** Mount the thread behind the hook eye, and wind the thread to the rear of the hook shank in close, tight wraps. Coat the thread wraps with flexible cement. Slide the foam up the hook to the tying thread; the longer tag of foam should be centered along the bottom of the hook shank. Take 3 thread wraps around the hook shank to secure the bottom foam strip. Then coat the inside of the segment with flexible cement.

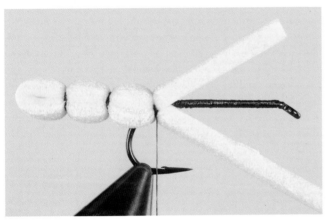

**8.** Place the top tag of foam over the hook shank, and secure it with 4 thread wraps.

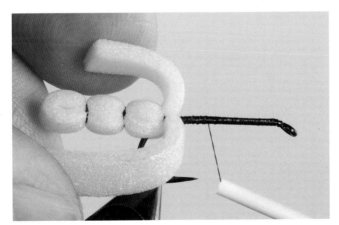

**9.** Fold the foam tags rearward, and wrap the thread forward one hook gap's distance.

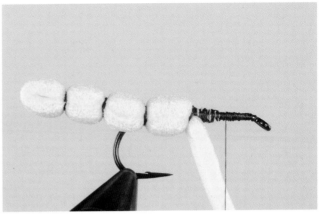

**10.** Form a fourth body segment as you did the third in steps 7–8. Then fold the bottom foam tag rearward, out of the tying field. Trim the tag from the strip of foam on top of the shank, and bind down the excess. Position the thread at the midpoint between the last segment and the hook eye.

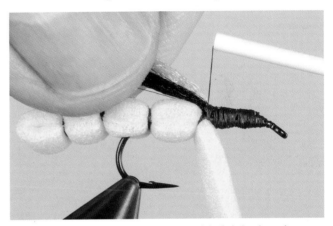

**11.** Mount a length of black yarn with 6 tight thread wraps. Then mount a length of tan yarn atop it. Trim the yarn tags, and bind them down. Coat the wing-mounting wraps with head cement. While holding the wings on top of the hook shank, wrap the thread rearward until it abuts the frontmost body segment, as shown.

**12.** Trim the wing so that it extends about $1/2$ of a hook gap beyond the body. Pull the bottom foam tag forward, center it snugly under the hook shank, and hold it in place. Then push a dubbing needle through the middle of the strip directly in front of the hook eye, as shown.

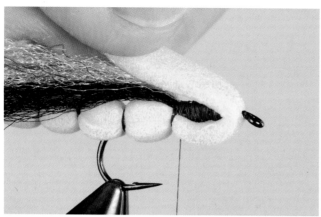

**13.** Coat the wing-mounting wraps with cement. Also apply cement to the inside face of the foam strip that will contact the wing-mounting wraps when the strip is folded over the top of the fly, as shown here. Pull the strip up over the hook eye, and gently work the hook eye through the hole formed in the foam strip. Fold the foam strip rearward over the top of the shank.

**14.** Secure the foam strip directly in front of the wing. Then trim the excess foam, leaving a short tag end.

**15.** With a light brown marker, color the bottom and sides of the body. Then apply thinned cement over the tinted portions of the body to lock in the color and prevent it from washing out during fishing.

**16.** Using the procedure described for the Green Drake Rubber Legs, steps 6–9, pp. 58–59, mount a length of rubber leg material on each side of the shank, as shown in this top view. Note that the front legs fit into the folds of the foam that forms the head of the fly. Trim the rear legs to be about even with the end of the body and the front legs to about the length of the hook shank. Secure the thread with 3 half hitches taken directly over the leg-mounting wraps, and trim the thread. Then check the alignment of the body, first by looking at the fly from underneath: the body should be directly in line with the hook shank, not angled or bent to one side. Now check the body by looking directly at the hook eye: the plane of the body should be horizontal, not twisted to one side. Since the glue is still pliable, bend or twist the body as necessary to get a straight, level body. Then coat the leg-mounting wraps with head cement. Apply cement to the foam that surrounds the hook eye to help prevent the foam from tearing. Set the fly aside to dry.

# Salmonfly

This largest of all the stoneflies gets its name from the salmon-colored bodies of the newly emerged adults, though within a day or two of hatching, this color changes to dark brown or black with orange bands and markings. Salmonfly nymphs take 2 to 4 years to mature and spend that time feeding among the rocks and boulders of well-oxygenated riffles and runs. Because of their preference for moderate to fast current flows, they are occasionally swept up into the drift. Streams with an abundance of these insects—particularly larger rivers, which tend to have denser populations—offer anglers the chance to catch trout any time of the year on nymph patterns. But from spring to early summer, when these insects hatch, is the prime time, and for the fish, the emergence of these insects could not be more opportune. Trout recovering from the lean winter months or the rigors of spawning gorge themselves on migrating nymphs and, when the chance presents itself, aggressively take any adult Salmonfly that floats near them. Hungry trout and the availability of meaty, protein-rich insects often combine to produce the most exciting dry-fly fishing of the season. Big fish rise readily, sometimes recklessly, and many dry-fly anglers take their largest fish of the season on adult stonefly imitations. There are, however, a couple drawbacks to this hatch:

the high, turbid waters of spring runoff can ruin the fishing, and when the rivers are in shape, the Salmonfly hatch attracts large numbers of fishermen. But despite the crowds, which are almost inevitable, this is an event that every angler should experience at least once. Go with an open mind, treat other anglers as you would want to be treated, and you'll have a good time. Think of this hatch more as a large social event, like a circus, with large trout as the star performers.

| | |
|---|---|
| **Other common names:** | Giant Salmonfly, Dark Stonefly, Black Stonefly |
| **Family:** | Pteronarcidae |
| **Genus:** | *Pteronarcys* |
| **Emergence:** | April to July, after dark |
| **Egg-laying flight:** | Late afternoon to dusk |
| **Body length:** | Mature nymph: 1–2 inches (25–50 mm); immature two-year-old nymph: ⅝–¾ inch (15–19 mm); immature one-year-old nymph: ⅜–½ inch (9–13 mm) |
| **Hook sizes:** | #4-8, 3XL for mature nymph; #6-10, 2XL for immature two-year-old nymph; #12-16, 2XL for immature one-year-old nymph |

**Above:** *Salmonfly nymph: dark brown to black.*

**Top right:** *Salmonfly nymph (underside): abdomen is lighter shade of top color; thorax is lighter shade of abdomen underside, with cream-colored gills and orange highlights.*

**Right:** *Salmonfly nymph (immature one-year-old): same colors as mature nymph.*

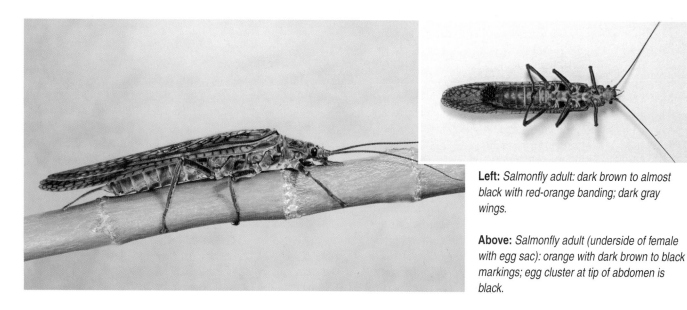

**Left:** *Salmonfly adult: dark brown to almost black with red-orange banding; dark gray wings.*

**Above:** *Salmonfly adult (underside of female with egg sac): orange with dark brown to black markings; egg cluster at tip of abdomen is black.*

# Important Fishing Stages

**Nymph** In the spring, as the water temperature nears 50 degrees F, the mature nymphs make their way to the shore during the late afternoon and evening; after dark, they crawl out of the water onto streamside rocks and vegetation, where the adults emerge. Though Salmonfly nymphs are available to trout all year, the days just before and during the hatch bring a mass migration of them toward the banks, and once they abandon the shelter of streambed rocks, many nymphs are swept into the current. Clumsy, ineffective swimmers, they drift in the flow until their own weight carries them back to the bottom. The days leading up to the emergence period and the first few days of the hatch, when trout are keying on the nymphs, are the most productive times to fish a large nymph imitation that matches the size of a mature natural. Dead-drift these big flies deep in current tongues, seams, runs, and riffles, particularly those near the banks. Multiple generations of Salmonfly nymphs inhabit the stream simultaneously, however, and once the hatch is over, the largest individuals are gone. During the rest of the year, smaller patterns better imitate the immature nymphs that remain. Fish these patterns close to the bottom with a drag-free drift in riffles and runs and the waters below them.

**Adult** After emerging, the adults gather on streamside foliage, often in great numbers, where mating takes place. As the temperature rises during the day, the adults become more active, clambering about on grasses, bushes, tree limbs, and often anglers. The fortunate result of all this activity among the adults is that many of them end up on the water, especially in windy weather. During the warm hours of the day, fish a dry fly in riffles and runs along the edges of the stream, concentrating particularly on areas where the current slides beneath overhanging foliage. From late afternoon to dusk, egg-laden females return to the riffles and runs almost anywhere across the breadth of the stream. While in flight, they briefly touch the water to release their eggs, and ultimately many of them fall spent to the water. These later hours of the day are the most productive times to fish a flush-floating dry fly in any likely-looking water.

# Salmonfly Nymph Patterns

**SHELLBACK SALMONFLY**

| | |
|---|---|
| **Hook:** | #4-8 4XL or 6XL nymph |
| **Weight:** | Lead or nontoxic wire |
| **Thread:** | Brown 3/0 or 6/0 |
| **Tail:** | Black goose or turkey biots |
| **Rib:** | Orange Vinyl Rib |
| **Shellback:** | Black Fino Skin or Thin Skin |
| **Body:** | Rusty brown 3-ply yarn |
| **Thorax rib:** | Midge Cream Cactus Chenille |
| **Legs:** | Black round rubber |

Though hardly an ultrarealistic imitation, this pattern emphasizes—even exaggerates—a few key components of the Salmonfly nymph that seem to play a role in the trout's recognition of the natural: the two-toned body; the dark, glossy wing case and segmented abdomen; the pronounced legs; and the cream-colored gills on the underside of the insect, which are not represented on most stonefly nymph patterns. Whether the gills contribute materially to the effectiveness of the fly is difficult to determine, but they are a distinctive feature of the pattern and at the very least show the trout something a little out of the ordinary. The use of yarn instead of dubbing for the body simplifies tying, but like most large patterns, this one does take a bit of time to dress. A very heavily weighted version, for deeper or faster water, is shown in the following sequence, but the pattern can be tied with less weight, in which case an extra layer of yarn over the thorax may be necessary to give the fly the proper proportions.

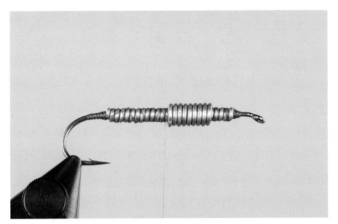

**1.** Using the instructions shown for the Pheasant Tail Nymph, steps 1–2, p. 21, wrap a layer of wire over the front ¾ of the hook shank, stopping 4 to 5 thread wraps' distance behind the eye. Secure the wire with thread. Wrap a second layer of wire over the first, beginning at the midpoint of the shank and ending 4 to 5 wire wraps' distance behind the frontmost wrap on the lower layer.

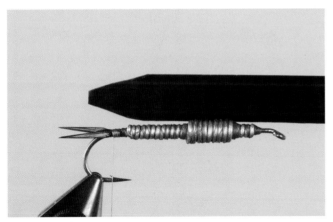

**2.** Secure the top layer of wire with thread wraps, and position the thread at the rear of the hook shank. Mount a biot on each side of the hook shank to form tails one hook gap in length as described for the Beadhead Prince Nymph, steps 4–5, p. 142. Clip and bind the excess biots, and position the thread at the rearmost mounting wrap. Cut a strip of the shellback material that is as wide as the hook gap and 1½ times the length of the hook shank. Taper one end as shown.

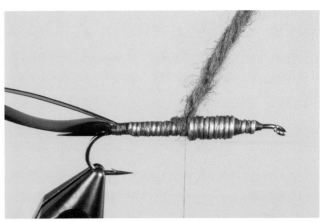

**3.** Mount the tapered end of the shellback material on top of the hook shank. Take 4 thread wraps forward, and mount the rib material on top of the hook shank. Wrap the thread forward, and mount the yarn directly behind the rearmost wrap of the top layer of wire.

**4.** With tight tension on the yarn, wrap it rearward; take one wrap behind the ribbing material, and wrap the yarn forward to the hanging thread, forming a smooth body.

**5.** Using your left fingers to support the hook as shown, wrap the yarn forward, ahead of the hanging thread, to a point just beyond the frontmost wrap of wire on the shank.

**6.** Wrap the yarn back to the hanging thread; secure the yarn and trim the excess. Draw the shellback material forward over the top of the body, and secure it at the hanging thread with 4 tight wraps. Spiral the rib forward, ending with one wrap taken ahead of the tag of shellback material. Secure the ribbing ahead of the shellback tag, and trim the excess.

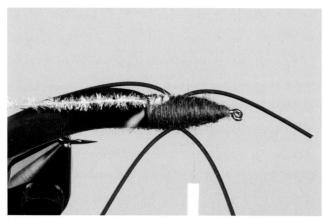

**7.** Mount the Cactus Chenille on top of the body directly in front of the shellback strip, as shown in this top view. Position the thread about $1/3$ the distance between the shellback mounting wraps and the hook eye. Mount the rubber legs on each side of the body as described for Green Drake Rubber Legs, steps 6–9, pp. 58–59. Position the thread halfway between the leg-mounting wraps and the hook eye.

**8.** Take one wrap of the Cactus Chenille directly in front of the shellback, another just behind the legs, one between the legs, and one in front of the legs just behind the hanging thread, as shown in this underside view.

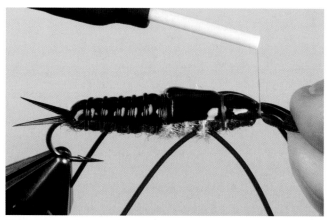

**9.** Secure the chenille on top of the body, and trim the excess. Position the thread about 3 thread wraps' distance in front of the chenille. Fold the shellback forward, and secure it with 4 tight thread wraps. Fold the shellback material rearward, and spiral the thread forward to the front of the body. Fold the shellback forward, and secure it behind the hook eye. Trim the excess, and finish the head. Trim the legs to $2/3$ the length of the body.

## GENERAL STONE

| Hook: | #4-8 2XH, 3XL–4XL nymph |
|---|---|
| **Weight:** | Lead or nontoxic wire |
| **Thread:** | Black 6/0 |
| **Tail:** | Black goose biots |
| **Abdomen:** | Orange vinyl ribbing |
| **Body:** | Black dubbing |
| **Wing case/head:** | Turkey feather sections, coated with Dave's Flexament |
| **Thorax:** | Peacock Ice Dub |
| **Legs:** | Black round rubber |

Year in and year out, this is probably our top-producing stonefly nymph pattern, particularly for fishing just before and during the hatch. The mature nymphs are large, and the conspicuous defining features—tails, legs, antennae, and wing cases-should be particularly visible to the trout. Perhaps this fly succeeds because it imitates these features a bit more realistically than simpler patterns do. But this is not an especially difficult fly to tie; it just takes a little time because of its large size. If there is any trick to dressing the fly, it's to make certain that the wing cases are mounted over a firm, tightly wrapped foundation of dubbing that is at least as large in diameter as the dubbed body behind it. Mounting the wing cases over a foundation that is too soft or small will cause them to flare unnaturally upward. Because this fly should be fished near the bottom, a heavily weighted version is shown here.

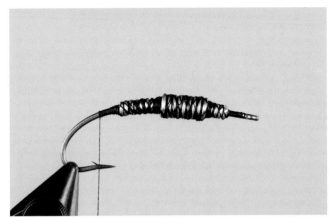

**1.** Using the instructions shown for the Pheasant Tail Nymph, steps 1–2, p. 21, wrap a layer of wire over the front ¾ of the hook shank, stopping 4 to 5 thread wraps' distance behind the eye. Secure the wire with thread. Wrap a second layer of wire over the first, beginning at the midpoint of the shank and ending 4 to 5 wire wraps' distance behind the frontmost wrap on the lower layer. Secure the second layer of wire with thread, and position the thread at the rear of the shank.

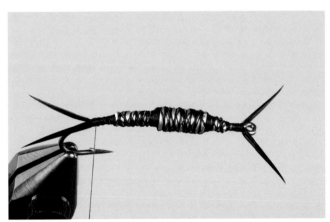

**2.** Mount a biot on each side of the shank to form tails one hook gap in length as described for the Beadhead Prince Nymph, steps 4–5, p. 142. Clip and bind the excess biots. Advance the thread to the hook eye, and mount 2 more biots to form antennae as long as the tails, as shown here from the top. Clip and bind the excess biots; position the thread about 5 thread wraps' distance in front of the rearmost wrap securing the tails.

**3.** Mount a length of vinyl ribbing material. Clip and bind the excess. Dub a slightly tapered abdomen over the rear ⅔ of the shank. Position the thread at the front of the abdomen.

**4.** Spiral the vinyl ribbing forward in 5 to 6 evenly spaced wraps. Tie off the rib in front of the abdomen, and trim the excess. Tightly dub a thorax just slightly larger than the thickness of the abdomen. Position the thread in the middle of the thorax.

**5.** Cut a strip of the turkey feather that is as wide as the hook gap. Fold it lengthwise, and trim one end at an angle; unfolded, the strip should have a V-shaped notch. Center the feather strip atop the thorax so that the tips of the V just reach the midpoint of the abdomen. Secure the strip with tight thread wraps.

**6.** Trim the tag of the wing-case feather, and secure with additional wraps. Mount a length of rubber leg material on each side of the shank directly atop the thread wraps used to secure the wing case, as described for the Green Drake Rubber Legs, steps 6–9, pp. 58–59. Trim the legs to the length of the abdomen, as shown in this top view.

**7.** Using a thinly dubbed thread, conceal the thread wraps used to mount the first wing case and the legs. Position the thread halfway between the leg-mounting wraps and the hook eye. Prepare a second strip of turkey feather identical to the first. Mount it atop the thorax so that the V tips extend rearward to the midpoint of the first wing case. Secure the feather strip tightly, wrapping forward to the hook eye, as shown in this top view. Do not clip the excess feather.

**8.** Dub rearward, forming a slightly tapered head that covers the thread wraps securing the second wing case. Position the thread at the rear of this dubbed head.

**9.** Fold the feather strip rearward over the top of the dubbed head, and secure it atop the shank. Finish the fly around these thread wraps. Trim the tag of the feather so that it reaches midway to the wing case behind it.

## Salmonfly Adult Patterns

**HENRY'S FORK SALMONFLY**

*Originator: Mike Lawson*

| | |
|---|---|
| **Hook:** | #6-8 3XL dry-fly |
| **Thread:** | Orange 3/0 or 6/0 |
| **Tail:** | Dark moose body hair |
| **Rib:** | Black or brown dry-fly saddle |
| **Body:** | Orange Antron yarn |
| **Wing:** | Natural dark elk hair |
| **Head/collar:** | Dark moose body hair |

Almost everything in this pattern conspires to provide good flotation—the buoyant moose and elk hair, the nonabsorbent body material, the hackle, and the bullet-head style. This is an excellent pattern for fishing rougher, more turbulent water, and the wing makes it visible even in a heavy chop. Surprisingly, though, this fly also can be effective on smoother bankside currents. Adult Salmonflies sometimes end up in the water and flutter their wings to propel themselves back to the shore, and the bushy, flared wing on this pattern may suggest to the trout the fluttering wings of the natural. Because the palmered body hackle is clipped, barb length is not especially important when choosing a feather, and this pattern can make good use of the long-barbed feathers on the butt end of a rooster neck, which are too large for most other dry-fly applications.

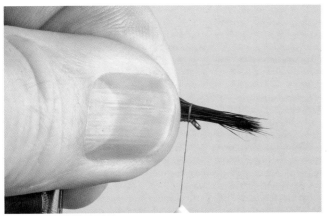

**1.** Mount the thread behind the hook eye, and wrap a tight foundation over the front ⅓ of the hook shank. Position the thread behind the hook eye. Clean and stack a bundle of moose body hair as shown in the instructions for the Sparkle Dun, steps 1–4, pp. 27–28. Position the hair atop the hook so that the tips extend ½ a shank length beyond the hook eye. Take 2 thread wraps around the bundle immediately behind the hook eye, using just enough tension to compress the hair slightly.

**2.** Pull downward on the bobbin to apply tension. As the thread pressure begins to roll the hair around the hook shank, continue wrapping over the hair bundle using tight thread tension. As the hair spins around the shank, try to keep the hair butts "caged" in your left fingers rather than releasing them. Keeping control of the hair butts here will simplify the next step. Take additional thread wraps around the bundle until the hair no longer spins around the shank.

**3.** Draw the hair butts rearward against the hook shank, and secure them with very tight thread wraps, stopping ⅓ of a shank length behind the hook eye. Trim away the butts of the hair, and wrap very tightly over the butts again back to the hook eye; make this thread foundation secure, since other materials will be mounted or tied off on top of it. Position the thread just behind the bound-down butts. Clean and stack a bundle of moose body hair. Position the bundle on top of the shank to form a tail one hook gap in length. Clip the base of the hair bundle so that it is even with the hair butts already bound to the shank.

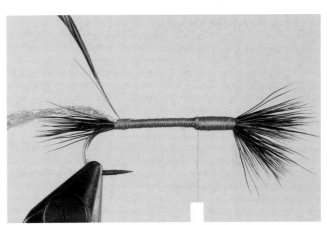

**4.** Mount the hair atop the shank, wrapping tightly rearward to the tailing point. Mount a length of Antron yarn atop the shank over the rearmost thread wrap securing the tail. Clip the excess yarn. Position the thread 4 thread wraps' distance ahead of the yarn-mounting point. Prepare and mount a hackle feather as described for the Little Green Drake, steps 4–6, p. 42. Position the thread ⅓ of a shank length behind the hook eye.

**5.** To form the body, take the first wrap of yarn directly abutting the rear of the hackle stem and the second wrap abutting the front of the stem. Then wrap the yarn forward to form a smooth, even body over the rear $^2$/$_3$ of the shank. Tie off and clip the yarn. Spiral the hackle forward in even turns.

**6.** Continue wrapping the hackle forward to the front of the body. Tie off the hackle as described for the Little Green Drake, steps 9–10, pp. 42–43. Trim away the feather tip. Position the thread at the front of the body. Working around the shank, clip the hackle barbs to $^1$/$_2$ the hook gap in length.

**7.** Clean and stack a bundle of elk hair. Position it atop the shank so that the hair tips extend to the end of the tail. Mount the bundle directly ahead of the body. Clip the excess hair, and bind down the butts. Position the thread at the rearmost thread wrap securing the wings.

**8.** Draw the hair tips that extend over the hook eye rearward. The hair tips should be evenly distributed around the hook shank, and the base of the hair should form a smooth, capsule-like head. Secure the hair at the base of the wings with a collar-like band of thread.

**9.** When the hair is secure, tie off the thread over the wraps used to form the head. Clip the thread. Trim away the hair tips that project below the hook shank.

## ROGUE FOAM GIANT STONE

*Originator: Jack Schlotter*

| | |
|---|---|
| **Hook:** | #4 TMC 200R |
| **Thread:** | Fluorescent fire orange 6/0 |
| **Body:** | Orange Larva Lace Dry Fly Foam, colored with dark brown marker |
| **Cement:** | Dave's Flexament, or similar flexible cement, thinned 1-to-1 with solvent |
| **Wing:** | Orange Krystal Flash, topped with a ⅜ x 4-inch strip cut from a plastic vacuum-sealing food bag |
| **Overwing:** | Moose body hair |
| **Head/collar:** | Dark natural or dyed-gray elk body hair |
| **Legs:** | Black round rubber |

An adult Salmonfly that ends up in the water may flutter its wings to help it get back to dry land, and fluttering stonefly patterns often work well. But far more often, an adult on the water runs shoreward across the surface or simply remains motionless on the film, which is especially the case with ovipositing females that have fallen spent to the water. Thus the most common profile of a drifting adult Salmonfly is with the wings folded along the back. We have experimented with a number of flatwing-style stonefly adults over the years, and this Jack Schlotter design is among the most productive and certainly the most durable. It's a particularly good choice for fishing smoother currents along vegetated banks during the daytime and for fishing anywhere during the evening when the females are laying eggs. However, one of its chief attributes—the low, flush-floating profile—is also one of its chief drawbacks: despite its size, this fly can be difficult to see in choppy or broken water. This pattern is a bit involved to tie, but the extra steps are worth the time, especially when it comes to heavily fished waters, where most other anglers are using bushy, fluttering stonefly designs. The wing on this pattern is formed from tinted and glued plastic film sheets; the assembly takes a few hours to dry. You may want to consult steps 13–14 before you tie this fly so that the wing material is ready when you sit down to the vise. Instead of the Larva Lace, you can substitute a strip of a similar ¹⁄₁₆-inch (2 mm) closed-cell foam. For information on types of foam suitable for this pattern, see p. 79.

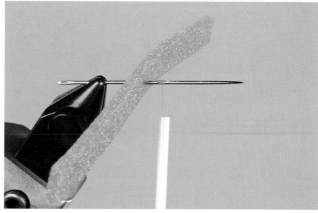

**1.** Place a needle in the vise. Take a 4-inch-long strip of ¼-inch-wide closed-cell foam, and push the very center of the strip onto the needle. Then mount the thread in front of the foam.

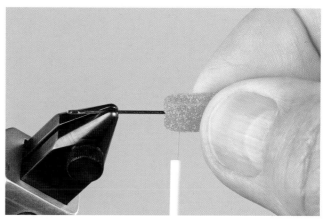

**2.** Fold the foam tags forward, sandwiching the needle between them. Slide the foam forward or rearward so that the distance between the fold at the rear and the hanging thread is ³⁄₁₆ inch.

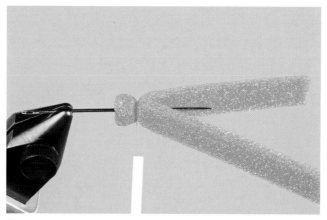

**3.** Take a loose wrap of thread around the foam, then pull firmly on the thread to seat it against the needle. Take 3 more tight wraps, and secure the wraps with 3 half hitches.

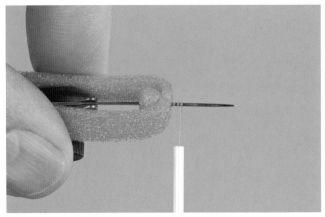

**4.** Fold the tags of foam rearward, and spiral the tying thread forward a distance equal to the width of the first segment, as shown in this top view.

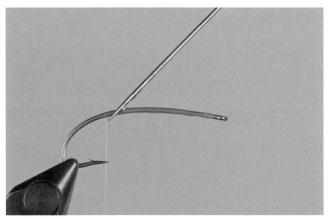

**7.** Mount the thread behind the hook eye, and wrap a thread foundation to the rear of the shank. Coat the thread wraps with cement.

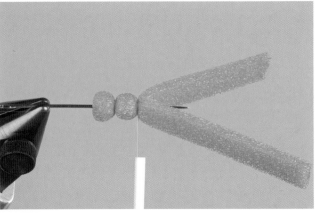

**5.** Fold the foam tags forward around the needle. Form the second body segment just as you did the first, and secure it with 3 half hitches.

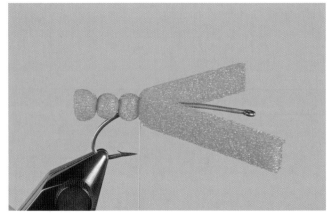

**8.** Sandwich the foam tags around the hook shank so that the thread is positioned to form a body segment equal in width to the segments on the extension. Secure the foam with 4 thread wraps.

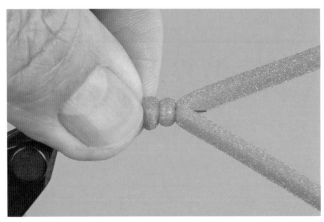

**6.** Clip the tying thread. Pinch the needle directly behind the rearmost segment, and push the foam off the needle. If any thread projects from the rear of the extension, clip it off. Then place a drop of cement on the thread wraps used to form each segment.

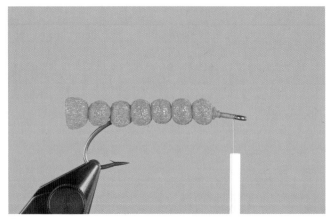

**9.** Form 4 more segments equal in size to the others. The frontmost segment should be about $5/16$ inch behind the hook eye. Clip and bind down the tag ends of the foam. Position the thread behind the hook eye.

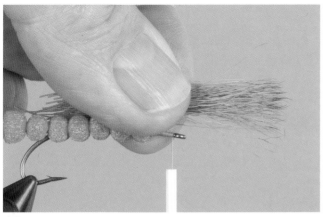

**10.** Clean and stack a bundle of moose body hair as shown in the instructions for the Sparkle Dun, steps 1–4, pp. 27–28. Position the bundle above the hanging thread so the tips extend about ½ a shank length beyond the hook eye.

**11.** Take one light thread wrap around the hair and hook shank. Take a second wrap, applying more tension to the thread to make the hair slip around the hook shank. After completing the second wrap, release the hair from your fingers. Add additional tight thread wraps directly over the first 2 wraps until the hair stops spinning. These wraps will distribute the hair evenly around the hook shank.

**12.** Trim the butt ends of the hair even with the front of the body, and bind them down. Mount 4 strands of orange Krystal Flash so that the tips extend to the rear of the body.

**13.** From the textured side of a vacuum-sealing food bag, cut two 2 x 4-inch sheets. Place the sheets, smooth sides down, on a piece of waxed paper. Color the top of one sheet with a black marker; coat the top of the other sheet with thinned flexible cement. Align the sheets, and press the inked side of the one against the glued side of the other. Then cover the sheets with a piece of waxed paper; place a book on top, and let the glue dry for a few hours.

**14.** From the inked sheet, cut a 2 x ³⁄₈-inch strip. Slightly taper the sides and round the corners of the wide end.

**15.** Position the tying thread behind the hook eye. Position the strip atop the hook shank so that the wing extends about the width of one body segment beyond the foam extension. Trim the front end of the wing just behind the hook eye. Mount the wing atop the shank by wrapping the thread tightly rearward to the front of the body.

**16.** Clean and stack a sparse bundle of moose hair. Mount it atop the shank so that the hair tips extend to the end of the wing. Clip and bind down the excess hair. Position the thread at the front of the body.

**17.** Draw the hair tips that extend over the hook eye rearward. The hair tips should be evenly distributed around the hook shank, and the base of the hair should form a smooth, capsule-like head. Secure the hair at the base of the wings with a collar-like band of thread. Trim away the hair tips that project beneath the hook shank.

**18.** Mount a length of rubber leg material on each side of the shank, as described for the Green Drake Rubber Legs, steps 6–9, pp. 58–59. Trim the front legs to one hook gap in length and the rear legs to 2 hook gaps in length, as shown in this bottom view. Tie off the thread with 3 half hitches around the leg-mounting wraps. Clip the thread. Use a dark brown marker to color the bottom half of the body. Coat the markings with thinned cement to lock in the color, and apply cement to the leg-mounting wraps.

# LOCALLY IMPORTANT HATCHES

## *Skwala*

Like most types of stoneflies, *Skwala* nymphs prefer the rocky or cobbled bottoms of riffles and fast runs, and some of the largest populations occur in lower-elevation trout streams. Even though it's not the first stonefly species of the year to emerge on most waters, many anglers regard the *Skwala* hatch as the beginning of their dry-fly season, and for good reason. First, though not a giant among stoneflies, the *Skwala* is still a sizable insect and makes a good meal for a trout during a comparatively lean time of the year. Second, after mating, the gravid females return to the stream and land on the water; they spend a good deal of time crawling over the surface laying their eggs, giving the trout ample opportunity to take them. And third, the hatch tends to occur before spring runoff, when the rivers are fishably low and clear, giving anglers a good opportunity as well. That having been said, *Skwala*s are rarely as prolific as Golden Stoneflies or Salmonflies; the hatches are more modest in scale, and the fishing, like winter and spring weather, can be spotty and change from hour to hour. Still, trout rising to big, visible dry flies—often for the first time in many months—make the *Skwala*s an early-season super hatch.

| | |
|---|---|
| **Other common name:** | American Springfly |
| **Family:** | Perlodidae |
| **Genus:** | *Skwala* |
| **Emergence:** | February to July, with March and April the prime months |
| **Egg-laying flight:** | Midday to late afternoon |
| **Body length:** | ¾–1¼ inches (20–32 mm) |
| **Hook sizes:** | #10-12, 2XL |

**Left:** Skwala *nymph: shades of brown with gold to olive mottling.*

**Above:** Skwala *nymph (underside): light brown to tan with yellow to olive tones.*

**Above:** *Female* Skwala *adult: body is brown to dark gray with yellow, olive, or orange mottling; dark dun wings.*

**Top right:** *Female* Skwala *adult (underside): brown and gold with yellow, olive, or orange tones.*

**Right:** *Male* Skwala *adult: shades of brown to dark gray with yellow, olive, or orange mottling; note the very short, nonfunctional, dark dun wings.*

# Important Fishing Stages

**Nymph** The nymphs are active predators and are occasionally swept into the current, where they drift for a distance before settling to the bottom again. Once they mature, the nymphs migrate from the faster flows to the margins of the stream; after dark, they crawl out of the water and onto streamside rocks or brush, where the adults emerge. During the migration, more nymphs are available to trout than at any other time of the year, but to call this preemergence behavior an "important" fishing stage may overstate the case a bit. Perhaps because *Skwala* hatches are seldom profuse, tending to the sparser side, the trout do not seem to key in on the nymphs as decidedly as they will during a Golden Stonefly or Salmonfly hatch. Fishing a *Skwala* nymph pattern close to the bottom in and below riffles or in currents near the shoreline, just before or during the hatch period, can certainly give results, but in our experience, a specific *Skwala* imitation isn't appreciably more productive than an impressionistic pattern of the proper size.

**Adult** Male *Skwala*s have very short wings and cannot fly. After emerging, they crawl along the shore searching for females, which they attract by drumming their abdomens, a behavior they share with other stonefly species. After mating takes place, the females conceal themselves in the streamside vegetation and wait for their eggs to mature. If disturbed, the adults hiding in the foliage typically drop to the ground and scurry for cover; if they land on the water, they crawl across the surface back to the shore. During the afternoon hours, the females return to the water to lay their eggs, either by crawling out onto the water from bankside cover or by flying out and landing on the surface; after laying their eggs, the females often continue moving about on the surface. Ordinarily, there are not large numbers of these adults on the water at the same time, but they are active and can bring trout to the surface. During this time, the angler's best bet is to fish to rises or cover likely holding waters. When there are no adults on the water, fish the edges of riffles and runs near the bank; trout move in quite close, waiting to capture the naturals.

# *Skwala* Nymph Pattern

## BH SQUIRREL NYMPH

*Originator: Dave Whitlock*

| | |
|---|---|
| **Hook:** | #10 2XL–3XL nymph |
| **Weight:** | Lead or nontoxic wire |
| **Head:** | ⅛-inch black or copper metal bead |
| **Thread:** | Black 6/0 |
| **Tail:** | Red fox squirrel guard hairs |
| **Rib:** | Medium oval gold tinsel |
| **Abdomen:** | ½ fox squirrel fur, ½ tan Antron dubbing, mixed |
| **Legs:** | Brown rubber leg material, medium |
| **Thorax:** | Red fox squirrel fur with guard hairs |

Trout seldom seem selective to *Skwala* nymphs, and in our experience, nymph-fishing the *Skwala* hatch is more like prospecting for trout than deliberately targeting a phase of the emergence. This variation of Dave Whitlock's Fox Squirrel Nymph is a reasonable representation of the size and color of the natural insect. It's quick and extremely simple to tie and has enough weight to get the fly deep, where it should be fished; the rubber legs give it good underwater movement.

**1.** Mount the bead on the hook, and wrap and secure a foundation of wire over the front half of the hook shank as described for the Beadhead Prince Nymph, steps 1–3, p. 141. Wrap the thread to the rear of the shank. Clip a bundle of fox squirrel hair from the hide. Pinch the tips of the long guard hairs, and remove the shorter underfur; set the underfur aside for dubbing. Mount the bundle of guard hairs to form a tail one hook gap in length. Clip and bind the excess. Position the thread at the rearmost tail-mounting wrap.

**2.** Mount a length of oval tinsel atop the rearmost tailing wrap. Clip and bind the excess. Position the thread at the rearmost tinsel-mounting wrap, and dub a slightly tapered abdomen over the rear half of the hook shank. Position the thread at the front of the abdomen.

**3.** Spiral the tinsel forward in the opposite direction from the one in which you wrapped the dubbing. Secure the ribbing in front of the abdomen; clip and bind the excess. Position the thread at the front of the abdomen. Dub a thorax, slightly larger in diameter than the abdomen, halfway to the rear of the bead. Position the thread directly ahead of the dubbing.

**4.** Mount a strand of rubber leg material on each side of the shank directly ahead of the dubbing, as described for the Green Drake Rubber Legs, steps 6–9, pp. 58–59, and shown in this top view.

**5.** Using a dubbed thread, conceal the leg-mounting wraps, and dub the remainder of the thorax forward to abut the rear of the bead. Tie off the fly behind the bead, and clip the thread. Trim the legs to the length of the body, as shown in this top view.

# *Skwala* Adult Patterns

**FOAM *SKWALA***

| | |
|---|---|
| **Hook:** | #8-10 standard dry-fly |
| **Thread:** | Yellow 6/0 |
| **Body:** | Tan Larva Lace Dry Fly Foam |
| **Cement:** | Dave's Flexament, or similar flexible cement, thinned 1-to-1 with solvent |
| **Wing:** | Tan over black Poly-Bear Fiber |
| **Legs:** | Small brown rubber leg material |

As is probably apparent, we favor this segmented foam abdomen for imitating larger insects. It creates a long body on a short hook and gives the impression of substantial body mass without a lot of weight or water-absorbent materials. The result is a flush-floating, buoyant pattern that is quite effective during the *Skwala* hatch. This variation of the Chugger-style fly that we use for larger stoneflies and grasshopper patterns omits the broad, folded-foam head to better represent the proportions of this smaller insect. Fish this fly dead-drift, with occasional twitches or a skittering, skating action to imitate these active insects. To form the body, you can either use a small strip of Pre-Sliced Larva Lace Dry Fly Foam or substitute a strip of a similar 3/32-inch (2 mm) closed-cell foam. For information on types of foam suitable for this pattern, see p. 79.

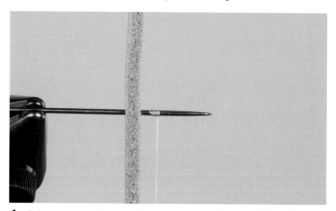

**1.** Mount a sewing needle in the vise. Take a 2½-inch-long strip of 1/8 x 3/32-inch (3 x 2 mm) closed-cell foam, and push the very center of the wider face of the strip on the needle. Mount the thread in front of the foam, and adjust the foam strip so that the hanging thread is 3/16 inch from the rear edge of the foam for a #10 hook (1/4 inch for a #8 hook), as shown in this top view.

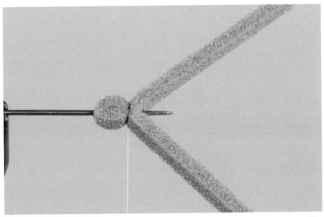

**2.** Fold the foam tags forward, sandwiching the needle between them. Take a loose wrap around the foam and gently tighten the wrap, as shown here from the top. Do not over-tighten the thread or you will cut through the foam. Take 4 more snug thread wraps directly over the previous wrap, then take 3 half hitches over the thread wraps to secure them.

**3.** Clip the thread, and place a drop of cement on the wraps. Pinch the needle tightly behind the foam, and push the foam off the needle.

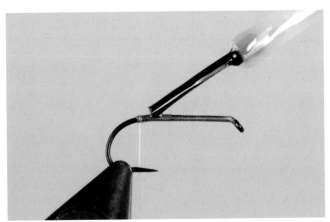

**4.** Place a hook in the vise. Mount the thread behind the eye. Using close, tight thread wraps, wrap the thread to the rear of the hook shank, and then forward 3 thread wraps' distance toward the hook eye. Coat the thread wraps with cement.

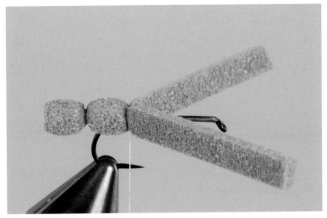

**5.** Sandwich the foam tags around the sides of the hook shank so that the thread is positioned to form a body segment that is the same size as the rear segment. Secure the foam with 6 snug thread wraps.

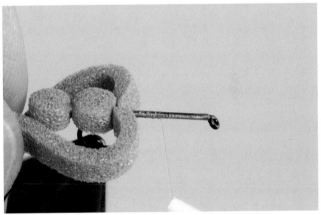

**6.** Fold the tag ends of the foam rearward, and spiral the thread forward a distance equal to the length of the previous segment.

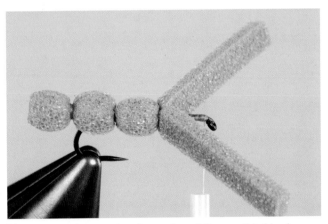

**7.** Fold the foam strips forward, and form a third body segment. Position the thread behind the hook eye.

**8.** Mount a small bundle of tan yarn atop the hook shank, extending over the hook eye. Mount a bundle of black yarn that is twice the thickness of the tan bundle on top of the tan yarn. Trim and bind down the excess yarn. Position the thread 4 to 5 thread wraps' distance behind the hook eye, and coat the thread wraps with cement.

**9.** Fold the foam strips forward, and secure them with 4 thread wraps; then take 3 half hitches to secure the thread. Pull the thread rearward over the top of the frontmost body segment, and take 2 thread wraps behind the first segment. Then trim the tags of foam even with the back of the hook eye; leave a short stub on each tag to keep the foam from slipping out from under the thread wraps. Place a drop of cement on the frontmost thread wraps.

**10.** Fold the black yarn rearward over the top of the body, and secure it with 2 tight thread wraps. Then fold the tan yarn rearward over the top of the black yarn, and secure it with 3 tight thread wraps. Trim the yarn wing so that it extends just beyond the rear of the body.

**11.** Using the instructions shown for the Green Drake Rubber Legs, steps 6–9, pp. 58–59, mount a strand of rubber leg material on each side of the body, as shown in this top view. Tie off the thread over the leg-mounting wraps with 3 half hitches, and cut the thread. Trim the legs to the length of the hook shank. After completing the pattern, while the cement is still pliable, view the fly from underneath and from the hook eye; if necessary, bend or rotate the body so that it is symmetrical from side to side on the hook and aligned along the axis of the shank. Apply cement to the leg-mounting wraps, and set the fly aside to let it dry.

## SKWALA MADAM X

*Originator: Doug Swisher*

| | |
|---|---|
| **Hook:** | #10-12 2XL dry-fly |
| **Thread:** | Yellow 6/0 |
| **Tail:** | Deer hair |
| **Rib:** | Tying thread |
| **Body:** | Deer hair |
| **Wing:** | Deer hair |
| **Legs:** | Small brown rubber leg material |

Many anglers use some version of the Madam X to fish other stonefly hatches, but the fly is particularly well suited to imitating the *Skwala* adult, which, compared with some other stonefly species, has a relatively slender body for its length. From a representational standpoint, this fairly minimalist pattern skeletonizes the imitation to emphasize a few key components in the trout's recognition of the natural insect: the long, low-floating body; the wings; and the highly mobile legs. This is an easy pattern to dress, and if there's any trick at all to the tying, it lies in accurately sizing the deer hair as shown in step 6 to ensure a wing of the proper length. When tying this fly on larger hooks, cut a few strands of deer hair from a patch of material, and size them against the shank to make certain the hair will be long enough to form the wing, since some deer hair is too short for the purpose. The pattern can be dressed with natural deer hair—the darker shades are a good match for *Skwala*s—or dyed-brown hair.

**1.** Mount the thread at the rear of the hook, and wrap a tight, uniform thread foundation over the rear 2/3 of the hook shank. Clean and stack a bundle of deer hair as shown in the sequence for the Sparkle Dun, steps 1–4, pp. 27–28. With your right fingers, position the bundle atop the shank so that the tips of the hair extend one hook gap's distance beyond the rearmost wrap of the thread foundation.

**2.** Gently push the hair bundle downward so that it surrounds the hook shank.

**3.** Maintaining the position of the bundle, transfer it to your left fingers, and mount the hair with 4 to 6 tight thread wraps. Trim the butt ends of the hair.

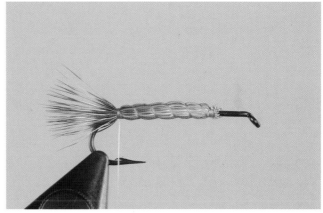

**4.** With your left fingers, draw the hair rearward so that it encloses the hook shank. Spiral the thread in evenly spaced turns to the rear of the thread foundation. Then take 4 tight wraps to secure the hair to the shank.

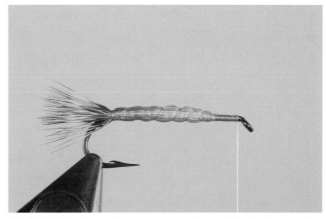

**5.** Spiral the thread forward to the front of the body, spacing the wraps the same distance apart as the wraps taken in step 4. Then wrap the thread tightly to the rear of the hook eye, binding down the hair butts that were left exposed in step 3. Position the thread behind the hook eye.

**6.** Clean and stack a bundle of deer hair. Size the bundle against the fly so that the hair extends from the end of the tail to the rear of the hook eye.

**7.** Mount the bundle of hair directly behind the hook eye, allowing the hair to slip around the hook shank under thread pressure. Secure the hair tightly with additional wraps. Trim the hair butts even with the front of the body, and bind down the butt ends. Position the thread at the front of the body, ⅓ of a shank length behind the eye.

**8.** Gather the hair in your left fingertips, and draw it smoothly and evenly rearward above the hook shank, forming a capsule-like head. Secure the hair with 5 to 6 tight thread wraps.

**9.** Mount a strand of rubber leg material on each side of the hook shank as described for the Green Drake Rubber Legs, steps 6–9, pp. 58–59. Tie off the thread directly atop the leg-mounting wraps. Clip the thread, and apply cement to the wraps.

**10.** Trim the rear legs even with the end of the tail and the front legs slightly shorter, as shown in this bottom view.

# Summer Stonefly

Summer Stonefly nymphs live in the colder, faster riffles and runs of larger trout streams in and around the Cascade and Rocky Mountains. Because these stoneflies emerge and lay their eggs after sunset, stay hidden during the day, and rarely hatch in great numbers, they often go unnoticed by anglers. In fact, only in comparatively recent years have Summer Stoneflies appeared with any real visibility in the fly-fishing literature. But they do leave noticeable clues during their emergence in the form of the discarded nymphal shucks attached to shoreline rocks, though anglers are apt to mistake these for the empty skins left by Golden Stoneflies and Salmonflies earlier in the season. The hatch generally lasts from 2 to 6 weeks, and on most streams the number of flies emerging each night is not large. Nevertheless, both the nymphs and adults are very active and are a significant food source for trout during the hatch cycle. The nymphs can be fished any time of the day. The adults are most active after dark, but as with many larger species of aquatic insects, trout continue to take their imitations during the day when no adults are present. In many waters, this hatch coincides with the appearance of grasshoppers, and though anglers may not be aware of the Summer Stoneflies, it's a pretty fair bet that many trout caught on hop-per patterns early in the day probably took the fly for a Summer Stone. You probably won't see these flies unless you go looking for them, since they conceal themselves during the daylight hours. The trout are quite aware of them, however, and knowledgeable anglers should be as well.

**Other common names:** Summer Stone, Short-Wing Stone, Tiger Stonefly
**Family:** Perlidae
**Genus:** *Claassenia*
**Species:** *sabulosa*
**Emergence:** June to September, after sunset
**Egg-laying flight:** After sunset
**Body length:** ¾–1¼ inches (20–32 mm)
**Hook sizes:** #6-8, 2XL

**Above:** *Summer Stonefly nymph: light to dark yellow-brown with dark brown markings.*

**Top right:** *Summer Stonefly nymph (underside): tan to cream.*

**Right:** *Summer Stonefly nymph: nymphal shucks are found very near the water's edge.*

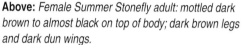

**Above:** *Female Summer Stonefly adult: mottled dark brown to almost black on top of body; dark brown legs and dark dun wings.*

**Top right:** *Female Summer Stonefly adult (underside): tan to cream.*

**Right:** *Male Summer Stonefly adult: mottled brown on top of body with lighter shade of top color on the underside; light brown legs and short wings.*

# Important Fishing Stages

**Nymph** These fairly large nymphs can take two to three years to mature. They are active predators, and in prowling about for food, they expose themselves to the current and are sometimes carried away in the drift. You can fish a nymph pattern close to the bottom in riffles and runs any time of the year, but certainly the most productive period is just before and during the hatch. Like the nymphs of most stonefly species, those of the Summer Stonefly migrate from faster water into the shallows when they are mature, and while crawling shoreward, they are particularly susceptible to being swept up in the current. Runs and riffles near or along the bank are worth searching with a nymph pattern, since the naturals continually move through these waters during the hatch period.

**Adult** The short-winged, flightless adult male Summer Stones actually look a good deal like the nymphs. But they are fast and agile runners, and you can sometimes see them scrambling away if you overturn rocks or debris on the banks. After emerging, the males scurry along the margins of the shore looking for newly emerged females with which to mate. Once mating has taken place, the females hide among the shoreline rocks and foliage and wait for their eggs to ripen. During the daytime, both the males and females remain concealed. After sunset, the males resume the search for emerging females, and the females, which are larger than the males and have fully developed wings, either fly or crawl to the water to lay their eggs. Adult Summer Stoneflies are very adept at scampering across the surface, and patterns that have lots of quivering rubber legs work particularly well at imitating these insects. Anglers do well to concentrate their efforts at the edges of riffles and runs along or near the banks; trout often lurk in the adjacent sheltered waters, waiting for the adult stoneflies to venture out onto the surface. Trout can hold quite close to the bank during this hatch, and wading anglers should take care not to walk through the water they should be fishing.

# Summer Stonefly Nymph Pattern

**STRETCH LACE NYMPH**

| | |
|---|---|
| **Hook:** | #6-8 2XL–3XL nymph |
| **Weight:** | Lead or nontoxic wire |
| **Thread:** | Tan or yellow 6/0 |
| **Tails:** | Spirit River Amber/Pepper Dancing Legs |
| **Abdomen:** | Small or medium tan or brown Stretch Lace |
| **Wing case:** | Brown ⅛-inch Scud Back |
| **Thorax:** | Tan dubbing |
| **Legs:** | Spirit River Amber/Pepper Dancing Legs |

The Stretch Lace cord used to form the abdomen on this pattern actually serves two purposes. First, it gives the fly a soft, almost glowing translucence that trout seem to find appealing. And second, the Stretch Lace produces a smooth, streamlined body that sinks more quickly, with less weight, than a shaggy, dubbed-body pattern, which has a higher degree of water resistance and thus requires more weight to carry it to the bottom. Less weight on the fly makes for more accurate and comfortable casting. Moreover, because Summer Stoneflies typically begin to hatch after runoff and may continue emerging late into the season when water levels are low, anglers may find themselves, at least some of the time, nymph-fishing fairly shallow flows. A fly that gets down quickly with a minimum of weight keeps the fly in the fishing zone longer while reducing the chance of snagging. Stretch Lace ties quite easily; the only real trick in tying it is taking care to form a smooth thread underbody with no ridges, thin spots, or abrupt changes in diameter. For a body with more golden color tones, use yellow thread for the underbody.

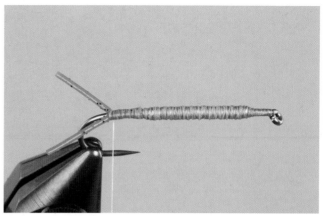

**1.** Wrap an underbody of wire over the middle ²/₃ of the hook shank as described for the Pheasant Tail Nymph, steps 1–2. p. 21. Mount the thread behind the eye. Form a tapered ramp of thread at each end of the wire underbody and secure the wire. Position the thread at the rear of the shank. Mount a length of rubber leg material on each side of the shank to form split tails about one hook gap in length. Clip and bind the excess.

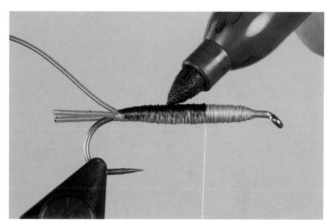

**2.** Mount a length of Stretch Lace very tightly atop the rearmost thread wrap securing the tails. Clip and bind the excess. Use the tying thread to form a slightly tapered thread underbody from the tail-mounting point to the frontmost wrap of wire. Work the thread back and forth to completely cover the wire and make the underbody smooth; a uniform foundation will give the best results when wrapping the Stretch Lace. Position the thread about ²/₅ of a shank length behind the hook eye. Color the top half of the abdomen with a brown marker.

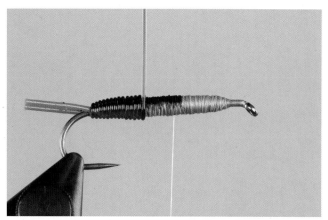

**3.** Pull the Stretch Lace firmly to stretch it, and begin wrapping it forward. As you wrap forward, gradually relax the tension on the strand. Beginning with tension on the material and progressively relaxing it assists in tapering the abdomen.

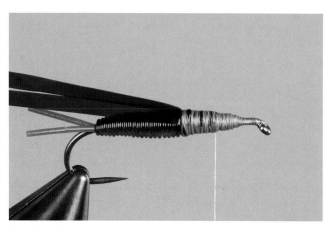

**4.** Continue wrapping to the hanging thread; secure the Stretch Lace, and clip the excess. Wrap the thread rearward to about the midpoint of the shank. Mount 2 lengths of Scud Back, side by side and slightly overlapping, directly ahead of the abdomen. Clip and bind the excess. Position the thread halfway between the front of the abdomen and the hook eye.

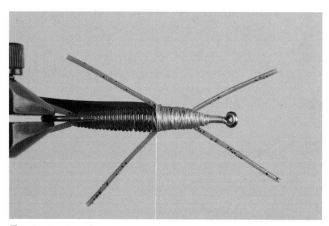

**5.** At the hanging thread, mount a strand of leg material on each side of the shank using the instructions shown for the Green Drake Rubber Legs, steps 6–9, pp. 58–59. When the strands are secure, draw the 2 back legs rearward, and wrap

the thread rearward, binding the legs against the sides of the shank, stopping just short of the abdomen, as shown in this bottom view. Clip the legs to about the length of the abdomen.

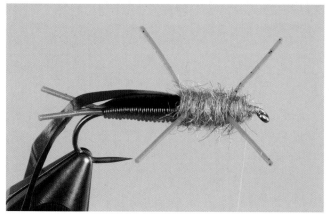

**6.** To form the thorax, take the first wrap or two of dubbed thread behind the rear legs. Then dub forward, taking a crisscross wrap of dubbed thread underneath the shank between the front legs to help conceal the leg-mounting wraps. Finish the thorax about 6 thread wraps' distance behind the hook eye.

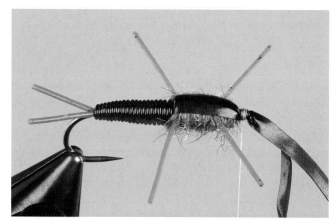

**7.** Draw the 2 strips of Scud Back over the top of the thorax; the strips should overlap slightly atop the thorax to give the impression of a single band of material. Secure the strips directly in front of the thorax. Clip and bind the excess. Finish the head of the fly.

# Summer Stonefly Adult Patterns

**MOJO**

| | |
|---|---|
| **Hook:** | #8-10 TMC 200R |
| **Thread:** | Tan 3/0 or 6/0 |
| **Body:** | Dark gold Antron dubbing |
| **Hackle:** | Brown |
| **Overbody:** | Tan 2 mm (¹⁄₁₆ inch) closed-cell foam |
| **Legs:** | Medium tan-and-black-barred rubber leg material |
| **Indicator:** | Orange and yellow 2 x 1 mm (¹⁄₁₆ x ¹⁄₃₂ inch) closed-cell foam |

This variation of the Chernobyl Ant was first shown to us by an angler on the South Fork of the Snake River in Idaho during the Summer Stonefly season. The silhouette, body colors, and segmentation provided by the hackle are all reasonable approximations of the natural, but the real key to this pattern is probably the abundance of long, wiggly rubber legs, which create movement on the surface, particularly if you give the fly an occasional twitch during the drift. The pattern can be tied without the indicator foam, but the fly floats low in the film, and even with the bright indicators, wading anglers, who are lower to the water than boating anglers, may have difficulty seeing the fly in choppy currents. The original version of this pattern used yellow-and-black barred legs, which is certainly an option. But the tan-and-black is somewhat closer color match to the naturals. This pattern has been a consistent producer during the Summer Stonefly hatch, particularly early in the day. Although we ordinarily use Larva Lace Dry Fly Foam for floating patterns, we prefer craft foam for the body of this fly. The denser, heavier craft foam is more appropriate for this pattern, which is designed to sit low in the water. For information on types of foam, see p. 79.

**1.** Mount the tying thread behind the hook eye, and wrap the thread rearward to a point ¹⁄₄ of a shank length behind the hook eye. Select a hackle feather with barbs about one hook gap in length. Prepare and mount the feather as described for the Little Green Drake, steps 4–6, p. 42. Wrapping rearward, dub a body that is slightly tapered at both ends; finish the body just above the hook point. Position the thread directly behind the body.

**2.** Spiral the hackle rearward over the body in 8 to 10 evenly spaced wraps, ending directly behind the body. Tie off the hackle as described in the sequence for the Little Green Drake, steps 9–10, pp. 42–43. Clip the feather tip. Take 3 tight thread wraps rearward, and wrap forward again to the last wrap of hackle. These thread wraps will furnish a foundation on which to mount the foam overbody. Cut a 1¹⁄₂- to 2-inch strip of foam that is one hook gap in width, and taper the rear end as shown.

**3.** Mount the tapered end of the foam strip atop the shank so that it extends rearward behind the body a distance of one hook gap. Secure the foam tightly.

**4.** Mount a length of rubber leg material on each side of the shank, using the instructions given for the Green Drake Rubber Legs, steps 6–9, pp. 58–59, shown in this top view.

**5.** Center a ¾-inch-long strip of the yellow indicator foam over the leg-mounting wraps. Secure the strip, and tie off the thread over the indicator-mounting wraps with 3 half hitches. Remount the tying thread behind the hook eye, and wrap rearward to the frontmost wrap of hackle.

**6.** Draw the foam strip forward over the top of the body, and secure it tightly to the top of the shank directly in front of the hackle. Clip the tag of foam extending over the hook eye to be about one hook gap in length.

**7.** Repeat steps 4–5 to mount the front legs and the orange indicator foam. Tie off the thread over the indicator-mounting wraps with 3 half hitches. Trim the front and rear indicators to about ½ of a hook gap in height.

**8.** Draw all the legs upward, and trim them to be as long as the hook shank. Use head cement to coat the thread wraps securing the overbody and indicator foam at the front and rear of the fly.

**9.** Trim the corners from the front tag of foam, as shown. Here's a bottom view of the finished fly.

## DIRTY-TAN CHUGGER

| | |
|---|---|
| **Hook:** | #8 Mustad 94840 |
| **Thread:** | Dark brown 6/0 |
| **Body:** | Tan Larva Lace Dry Fly Foam, tinted with brown marker |
| **Cement:** | Dave's Flexament, or similar flexible cement, thinned 1-to-1 with solvent |
| **Wing:** | Tan over black Poly-Bear Fiber |
| **Legs:** | Brown rubber leg material |

While this pattern floats especially well in broken or choppy water, one of its big advantages is its visibility to the angler. Even though Summer Stones don't tend to flutter on the water as much as other stonefly species, the wing profile on this pattern doesn't appear to be a particular drawback in the effectiveness of this fly, provided that the wing is not so heavily dressed that it angles upward dramatically. This fly is essentially identical to the Dirty-Yellow Chugger and is tied using the instructions shown on pp. 197–200, with one difference: for this pattern, form only a single body segment on the needle; then remove the foam extension, and mount it on the hook. Instead of the Larva Lace, a strip of a similar $1/16$-inch (2 mm) closed-cell foam can be substituted. For information on types of foam suitable for this pattern, see p. 79.

# Yellow Sally

The nymphs of Yellow Sally stoneflies have no external gills, so they require well-oxygenated water like that typically found in the riffles and runs of cold mountain streams. But larger lowland rivers that remain cool during the summer months can also have fishable populations of these insects in sections with riffly or broken water, which has a higher dissolved-oxygen content. The numbers of these stoneflies and the densities of the hatches vary from stream to stream and even from year to year on a given river, but the trout feed readily on Yellow Sallies when an emergence occurs. It's not difficult to recognize a hatch of these insects. The light colored adults fly above the water with the vaguely prehistoric profile characteristic of all airborne stoneflies, and on the banks and streamside vegetation, they are active runners. Yellow Sallies are distributed widely throughout the West, and because of their habitat requirements, they are often one of the better hatches in the headwaters of streams.

| | |
|---|---|
| **Other common names:** | Little Yellow Stonefly, Stripe Tail |
| **Family:** | Perlodidae |
| **Genus:** | *Isoperla* |
| **Emergence:** | May to October, with the best fishable hatches from June to August, late afternoon to dark |
| **Egg-laying flight:** | Midafternoon to dusk |
| **Body length:** | ¼–¾ inch (7–19 mm) |
| **Hook sizes:** | #8-16 |

**Above:** *Yellow Sally nymph: light yellow to brown with dark markings on back.*

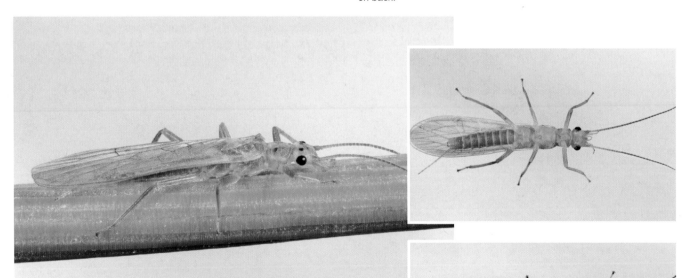

**Above:** *Yellow Sally adult: gold to yellow.*

**Top right:** *Yellow Sally adult (underside): some insects show a reddish orange tint on the rear abdominal segments.*

**Right:** *Yellow Sally adult (underside): other insects lack the reddish orange tint.*

## Important Fishing Stages

**Nymph** These predaceous nymphs are quite active as they prowl the streambed for food, and they are occasionally swept into the current. But unless a stream has a large population of Yellow Sallies, such mishaps do not produce noteworthy angling opportunities, and the best time to fish a nymph pattern is when the adult flies are seen along the stream during a hatch period. What anglers call Yellow Sallies are really a group of stoneflies that are similar in appearance. Some of the species in this group emerge in open water, in the riffles or areas below them, and when this type of hatch is occurring, fish an unweighted nymph pattern as a dropper below a dry fly, such as an adult Yellow Sally pattern. Most species in this group of stoneflies, however, behave like their larger cousins, migrating from the faster currents to the sheltered margins of the stream and crawling onto the shore between late afternoon and dusk to emerge. During this migration, the nymphs are both more concentrated and more vulnerable to the force of the current; fishing a nymph imitation close to the bottom, particularly in currents near the shoreline, is a productive approach.

**Adult** After emerging, the adults gather in streamside foliage, where they mate, and the females wait until their eggs have matured. Egg-laden females return to the water, gathering in the air over riffles from midafternoon to dusk. They drop to the surface for a few moments to release clusters of eggs, and then fly back into the air and rejoin the swarm. They will continue dropping to the water and lifting off again until the last of their eggs are released, at which point they lie spent on the water. The constant activity of even a small number of females repeatedly hitting the water, laying their eggs, and flying upward brings trout to the surface. Rises can be a bit difficult to spot in the choppier riffles or in low light late in the day, but for the most part, the splashy rises that these insects induce are hard to miss. During egg-laying flights, fish a dry fly by itself or use a dropper below it—either an unweighted nymph pattern, if the insects are emerging in open water, or a soft-hackle pattern that imitates a drowned adult.

## Yellow Sally Nymph Patterns

**BIRD'S NEST**

*Originator: Cal Bird*

| | |
|---|---|
| **Hook:** | #8-14 2XL nymph |
| **Weight (optional):** | Lead or nontoxic wire |
| **Thread:** | Tan 6/0 or 8/0 |
| **Tail:** | Wood-duck flank-feather barbs |
| **Rib:** | Gold or copper wire |
| **Abdomen:** | Tan or light brown Australian opossum fur |
| **Hackle:** | Wood-duck flank feather |
| **Thorax:** | Tan or light brown Australian opossum fur |

The Bird's Nest is often fished as a general attractor nymph, but the body proportion, shape, and coloration suggest a Yellow Sally nymph. And as often happens in fly fishing, more generally impressionistic patterns with a few basic correlations to the natural insect frequently outfish more scrupulously representational types. This fly is tied in the round; that is, it is symmetrical around the axis of the hook shank and has no top or bottom. Tumbling in the current, it always presents the same broadside silhouette to the trout. The pattern can be tied weighted for fishing deep or unweighted for fishing higher in the water column to imitate species that emerge in open water. This fly can be hackled in a number of different ways, from wrapping a feather to using stripped barbs distributed around the shank, as shown in the following sequence. Regardless of the method, the hackle should be kept fairly sparse; the pattern should look more like a nymph than a heavily hackled wet fly. Instead of wood duck, you can substitute teal flank feathers.

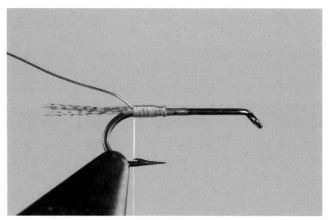

**1.** Mount the thread at the midpoint of the hook, and wrap to the rear of the shank. Align a section of 10 to 12 wood-duck flank-feather barbs, and strip them from the stem. Mount them at the rear of the shank to produce a tail about ½ a shank length long. Clip and bind the excess, and return the thread to the rearmost tail-mounting wrap. Mount a length of ribbing wire, and bind the excess. Position the thread at the rear of the shank.

**2.** Dub a slightly tapered body over the rear ²/₃ of the shank, and position the thread at the front of the body. Spiral the ribbing wire forward over the abdomen in evenly spaced turns.

**3.** Spiral the ribbing wire to the front of the abdomen. Tie it off in front of the abdomen; clip and bind the excess. Position the thread halfway between the abdomen and the hook eye.

Align a section of wood-duck barbs that are at least slightly longer than the hook shank, and clip them from the stem. Position the bundle of barbs atop the shank so that the tips reach the hook bend.

**4.** Maintaining the position of the bundle, transfer it to your left fingers. Using moderate thread tension, begin wrapping over the bundle. As the thread contacts the barbs, do not wrap over them, but instead use the thread pressure to push some of the barbs down the far side of the shank. As the fibers begin to move down the far side, continue wrapping and use your left fingers to roll the barbs farther around the shank. Continue the wrap, pushing the barbs around the shank with a combination of thread pressure and help from your fingers. The object here is to get a layer of barbs distributed evenly around the perimeter of the hook shank.

**5.** After you have taken the first wrap, release the barbs from your left fingers, and take a tight thread wrap toward the hook bend. This wrap will help further distribute the barbs. Once the barbs are deployed fairly evenly, wrap the thread rearward, tightly abutting the front of the abdomen to help flare the barbs outward. Position the thread at the front of the abdomen.

**6.** Clip the excess feather barbs. Dub a thorax that tapers toward the hook eye. Tie off the thread behind the hook eye. Here's a front view of the finished fly.

## PARTRIDGE AND YELLOW

| | |
|---|---|
| **Hook:** | #10-14 2XL nymph |
| **Thread:** | Yellow 8/0 |
| **Abdomen:** | Yellow floss |
| **Thorax:** | Hare's ear dubbing |
| **Hackle:** | Gray or brown partridge |

Soft-hackle patterns are most often fished with a tight line on the swing, but they can also be highly effective on a dead drift—on, in, or just beneath the surface film. Fished drag-free, this Partridge and Yellow credibly represents a spent or drowned adult; the soft, mobile hackle suggests the general dishevelment of wings and legs on a dead or dying adult. It's also an effective pattern for imitating the emerging adults of those species that hatch in open water. Either way, it can be a difficult fly to see, particularly when fished awash in the film or just below the surface. Hookups are much improved if the fly is fished as a dropper behind an adult Yellow Sally pattern. The key in dressing this pattern, as with soft-hackle designs generally, is to keep the hackle sparse.

**1.** Mount the thread behind the hook eye, and position it about 4 thread wraps' distance behind the eye. Select a hackle feather. Locate the lowermost barbs that are twice the hook gap in length. Strip away all the barbs below this point.

**2.** Mount the feather atop the shank, as shown, with the dull or concave side facing you. Flatten the thread by spinning the bobbin counterclockwise (when viewed from above). Bind the feather stem by wrapping rearward to the midpoint of the hook shank, forming a smooth thread foundation as you wrap. Clip the feather stem at the midpoint of the shank.

**3.** Wrap the flattened thread forward to a point 5 thread wraps' distance behind the mounted hackle. Position a length of floss atop the shank, and take 2 firm but not tight thread wraps over the floss. Then pull the floss rearward so that it slides beneath the thread wraps, leaving a short tag. Secure the tag with 2 tight wraps. Sliding the floss beneath the wraps helps align the floss filaments to simplify wrapping and produce a neater body.

**4.** Draw the floss beneath the shank to form a smooth band of material. Wrap to the rear of the shank in overlapping turns. Then wrap forward again, gradually increasing the amount of overlap in the wraps to give the body a slight taper. When you reach the tying-thread position, secure the floss.

**5.** Clip and bind the excess floss. Dub a thorax forward, ending 2 thread wraps' distance behind the mounted hackle. Position the thread in front of the thorax.

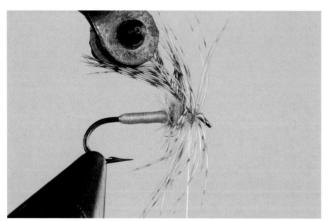

**6.** Clip the feather in a pair of hackle pliers, and raise it vertically so that the front or convex side of the feather faces the hook eye. Take 2 wraps of the feather rearward toward the thorax.

**7.** Take one wrap of thread over the feather tip, directly in front of the thorax.

**8.** Take 2 thread wraps forward through the splayed hackle barbs, ending in front of the first hackle wrap. Wiggle the thread as you wrap forward to avoid trapping any of the barbs beneath the thread. Trim away the feather tip, and finish the fly behind the hook eye.

**9.** Here's a front view of the finished fly.

# Yellow Sally Adult Patterns

## ELK HAIR CADDIS

*Originator: Al Troth*

| | |
|---|---|
| **Hook:** | #10-16 standard dry-fly |
| **Thread:** | Tan 6/0 |
| **Rib:** | Fine gold wire |
| **Body:** | Yellow or tan dubbing |
| **Wing:** | Light elk hair |
| **Hackle:** | Ginger or light dun dry-fly |

Although this pattern is widely used to imitate lighter-colored species of caddisflies, it makes an excellent adult Yellow Sally imitation when dressed with a yellow or tan body. The light colored materials are a good match with the naturals, and it has the added advantage of being a very congenial fly to the fisherman, particularly in the riffly or choppy water where egg-laying females are commonly encountered. The fly floats well and can easily support a soft-hackle or Yellow Sally nymph imitation as a dropper; it is quite visible to the angler; and owing to the palmered hackle, it skitters well on the surface, which can sometimes be an effective technique during the hatch. Many anglers are convinced that the effectiveness of an adult Yellow Sally imitation is improved by adding a bit of red color to the end of the abdomen to represent the egg sac of an ovipositing female. We can't really say that this has held true in our own experience but don't doubt that it can make a difference on some waters or at certain times. And it is no large matter to add a touch of red dubbing at the rear of the abdomen.

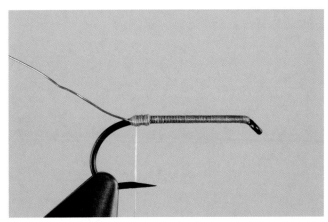

**1.** Mount the thread behind the hook eye, and wrap it to the rear of the shank. Mount a length of ribbing wire on the near side of the hook shank. Clip and bind the excess, and position the thread at the rearmost wire-mounting wrap.

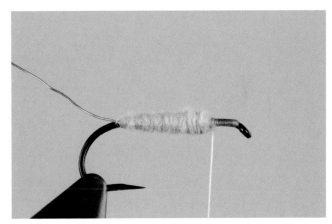

**2.** Dub a slightly tapered body over the rear ¾ of the shank, and position the thread at the front of the body.

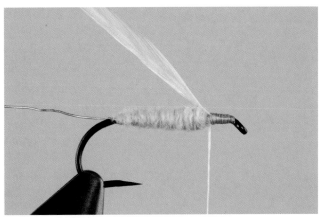

**3.** Select a hackle feather with barbs about 1½ hook gaps in length. Prepare the feather and mount it directly ahead of the abdomen as described for the Little Green Drake, steps 4–6, p. 42.

**4.** Spiral the feather rearward to the end of the abdomen in evenly spaced wraps. With your left fingers, hold the tip of the feather above the shank and slanted slightly rearward. With your right fingers, take a complete wrap of the ribbing wire around the feather tip, ending with the wire held above the shank and angled slightly forward.

**5.** Spiral the ribbing wire forward, wiggling it slightly from side to side as you wrap in order to prevent the rib from binding down any hackle barbs.

**6.** When you reach the front of the abdomen, secure the wire and clip the excess. Position the thread 2 to 3 thread wraps' distance ahead of the abdomen. Then trim off the hackle tip at the rear of the fly. Clean and stack a bundle of elk hair as described for the Sparkle Dun, steps 1–4, pp. 27–28. Position

the bundle atop the shank so that the hair beyond your fingertips extends from the hook eye to the hook bend. The mounting point for the hair is at the right fingertips.

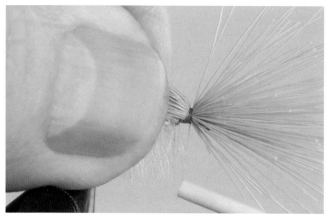

**7.** Transfer the bundle to your left fingers, and mount the bundle atop the shank with 6 to 8 tight thread wraps.

**8.** Lift the butt ends of the hair, and take 4 to 5 thread wraps tightly abutting the base of the hair.

**9.** Finish the fly around the wing-mounting wraps, and clip the thread. Trim the butts of the hair at a slant that follows the angle of the hook eye.

## HENRY'S FORK YELLOW SALLY

*Originator: Mike Lawson*

| | |
|---|---|
| **Hook:** | #12-16 2XL dry-fly |
| **Thread:** | Yellow 6/0 |
| **Abdomen:** | Yellow dubbing |
| **Wing:** | Light deer hair |
| **Hackle:** | Light dun dry-fly |
| **Thorax:** | Yellow dubbing |

This simple fly is a productive imitation of an egg-laying or spent Yellow Sally. Though the general design is similar to that of the Elk Hair Caddis, its characteristics on the water are somewhat different. With the fairly sparse wing and modest number of hackle wraps, this is a low-floating fly, and it's wise to keep it well dressed with floatant if fishing choppy water. But the uncluttered dressing and low-riding body that create a well-defined silhouette when viewed from underneath are partly what account for its effectiveness. In fact, some anglers trim away the hackle barbs beneath the hook shank, as shown in step 7, to put the body right against the film. This pattern is especially well suited to the flatter waters below riffles, where trout hold to collect spent flies. When dressing this pattern, choose a fine-textured, synthetic dubbing such as superfine poly, which dubs tightly to create a smooth, sharp body profile. We like light colored deer hair for the wing, but elk with light tan or pale gold tips can also be used.

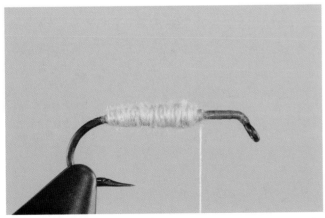

**1.** Mount the thread behind the hook eye, and wrap to the rear of the shank. Dub a slightly tapered abdomen over the rear ²/₃ of the shank. Position the thread directly in front of the abdomen.

**2.** Clean and stack a bundle of deer hair as described for the Sparkle Dun, steps 1–4, pp. 27–28. Position the bundle atop the shank so that the hair tips extend to the hook bend.

**3.** Mount the hair bundle directly in front of the abdomen with 5 to 7 tight thread wraps.

**4.** Trim the hair butts at an angle tapering toward the hook eye. Bind the butts down, forming a smooth thread foundation as you wrap. Position the thread at the rearmost wing-mounting wrap. Prepare a hackle feather and mount it directly ahead of the abdomen as described for the Little Green Drake, steps 4–6, p. 42. Position the thread at the rearmost hackle-mounting wrap.

**5.** Dub the thorax forward, ending about 4 thread wraps' distance behind the hook eye. Position the thread at the front of the thorax.

**6.** Spiral the hackle forward in 3 to 4 wraps to the front of the thorax. Tie off the hackle as described for the Little Green Drake, steps 9–10, pp. 42–43. Finish the head of the fly.

**7.** Here's a front view of the finished fly with the hackle barbs beneath the shank clipped away.

# Chapter Four

# Midges

Though the term "midge" is often used indiscriminately to describe any tiny, winged insect—from no-see-ums to gnats to mosquitoes—among anglers, the term designates the small, nonbiting, two-winged adults from the family Chironomidae. With more than 175 genera and 1,000 species, this is a particularly large and widespread family of aquatic insects. Midge larvae are found in most still and moving waters, though as a rule, those that live in lakes tend to be larger than those in rivers. Midge populations vary from stream to stream, but the slower, more consistent flows of tailwaters and spring creeks often host the most abundant populations. Most stream midges produce multiple generations annually, and peak emergence periods occur throughout the year in ice-free waters. Although midges are small insects—sometimes extremely small—their aggregate biomass can be substantial; the larvae, pupae, and adults constitute a highly important food source for trout, especially during the fall, winter, and spring months, when few other aquatic insects are available in large numbers.

| | |
|---|---|
| **Other common name:** | Chironomid |
| **Family:** | Chironomidae |
| **Emergence:** | All year in ice-free waters |
| **Egg-laying flight:** | Throughout the day |
| **Body length:** | 1/16–3/4 inch (2–20 mm); in streams, most are 1/16–1/4 inch (2–7 mm) |
| **Hook sizes:** | #16-22 |

**Above:** *Midge pupa: shades of brown, green, tan, cream, red, or black.*

**Above:** *Midge larva: shades of tan, brown, green, cream, or black.*

**Left:** *Bloodworm: these bright red midge larvae contain hemoglobin, which allows them to store oxygen in their bodies.*

237

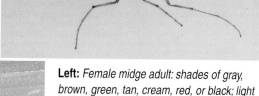

**Left:** *Female midge adult: shades of gray, brown, green, tan, cream, red, or black; light dun wings.*

**Above:** *Female midge adult (underside): same as top color.*

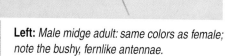

**Left:** *Male midge adult: same colors as female; note the bushy, fernlike antennae.*

**Above:** *Male midge adult (underside): same as top color.*

# Important Fishing Stages

**Larva** Midge larvae live on or burrow into the substrate, and streambeds that accumulate debris tend to have the largest populations. Midge larvae can be found tumbling in the current all year round, but greater numbers of them appear in the drift when the stream bottom is disturbed by spates, the strong flows of high water, or even wading anglers. At one time, in fact, some anglers would intentionally kick up debris on the stream bottom to dislodge midge larvae and attract trout to the water downstream—an illegal activity known as the "San Juan shuffle" for the river where it was once a common occurrence. You can fish a larva pattern just off the bottom in riffles, runs, and pools at any time of the day or year and catch trout. Even during emergence periods for other insects, trout often take a midge larva pattern fished as a dropper behind a larger nymph. Although midge larvae are poor swimmers, they wiggle around a good bit when drifting freely in the current. Using a small loop knot to secure the fly to the tippet allows a larva pattern to swing about in the current, imitating the movement

of the naturals. The large number of midge species can pose a problem when fishing, since the various larvae can differ significantly in size and color. Unless you can capture a natural larva of the prevailing species for the location or time of year, fly selection is a hit-or-miss proposition. Size 18 is a good opening gambit, and dark brown, black, green, and red are the most common body colors. When fishing, periodically change colors until you start catching trout.

**Pupa/emerger** Midge larvae pupate on the bottom of the stream. At emergence, they slowly swim and drift upward until they break through the surface film, where the adults emerge. Midges become more vulnerable to trout during this pupal stage than in any other phase of the life cycle. They are completely exposed, swim weakly, and typically drift for long distances before reaching the surface. A pupa pattern can be fished at any time of the day and at all levels of the water column, but ordinarily the two best zones to fish are the bands of water within a foot of the streambed and within a foot of the surface. If you see midge adults on or over the water but no surface activity, fish a pupa pattern close to the bottom, by itself or trailing behind a larger nymph. If you see midge adults and trout feeding close to the surface, fish a pupa pattern on an 8- to 12-inch dropper below a small indicator or dry fly. When trout are feeding from the surface or just below the film, use a floating emerger pattern.

**Adult** After a pupa breaks through the surface film, it generally spends very little time on the water before flying off, though in colder weather, the amount of time spent on the surface increases. During a midge emergence, trout tend to feed on the pupae more than the adults, but fish taking pupae close to the surface will often take a floating adult as well. An adult midge pattern with a pupa pattern trailing behind it is a very effective combination to fish during an emergence. Mating takes place in swarms on or above the surface of the water. After mating, the females land on almost anything projecting above the water—sticks, rocks, aquatic vegetation, even anglers—and then crawl under the surface to lay their eggs. Afterward, the spent females usually end up drifting in the current. Trout feed on the mating swarms that congregate on the surface as well as on the spent females. A cluster midge pattern is used to represent a midge swarm, and a sunken-adult pattern imitates a spent female. Fishing these patterns in tandem, with a midge cluster as a dry-fly/indicator and a sunken adult as a dropper, is a productive approach during a midge hatch.

# Midge Larva Patterns

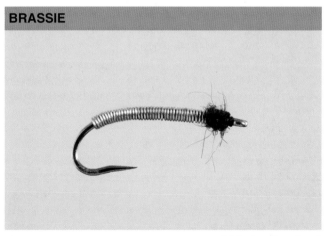

**BRASSIE**

*Originator: Gene Lynch*

| | |
|---|---|
| **Hook:** | #16-22 2XL–3XL nymph |
| **Thread:** | Black 8/0 |
| **Body:** | Copper wire |
| **Head:** | Dark brown or black dubbing |

From a fly-tying standpoint, the segmented, wormlike midge larva is a pretty basic package, and the patterns used to imitate it, like the Brassie, tend to be rather simple in design—further proof that a fly need not be complicated to take fish reliably. The effectiveness of this pattern probably owes as much to the material as to the appearance. The copper wire body adds a fair amount of weight to a small hook,

and this fly readily penetrates the surface film and sinks relatively quickly, making the pattern a good choice for faster water. To keep the body from being unnaturally plump and to facilitate tying, use copper wire with a diameter no larger than that of the hook wire. For all those reasons why fly fishing continues to be a mystery, natural copper wire makes a productive pattern, even though we have never seen a midge larva of a similar color. But copper wire now comes in a variety of colors, and we also tie this fly in red, brown, and green to match common midge colors.

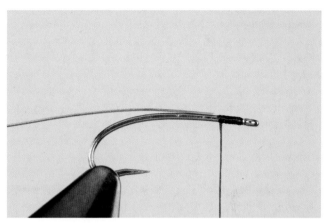

**1.** Mount the thread behind the hook eye, and position it 6 to 7 thread wraps' distance behind the eye. Position a length of copper wire atop the shank so that the end of the wire is at the hanging thread.

**2.** Flatten the thread by spinning the bobbin counterclockwise (when viewed from above), and secure the wire to the rear of the hook shank.

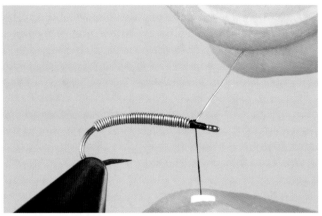

**3.** Position the thread 6 to 7 thread wraps' distance behind the eye. Wrap the wire forward in smooth, touching turns up to the hanging thread. Secure the wire atop the shank with 4 to 5 tight wraps. Use your left hand to put some tension on the bobbin, and wiggle the wire back and forth with your right fingers until it breaks.

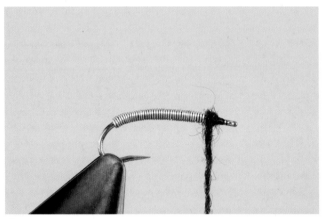

**4.** Secure the wire tag with thread. Wrap the thread back over the wire body to a point about $\frac{1}{5}$ to $\frac{1}{6}$ of a shank length behind the hook eye. With a dubbed thread, form a thorax that is about twice the thickness of the body. Finish the head of the fly.

## KRYSTAL FLASH MIDGE

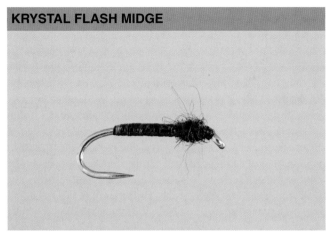

| | |
|---|---|
| **Hook:** | #16-20 1XL–2XL nymph or scud |
| **Thread:** | Dark brown 8/0 |
| **Body:** | Black or green Krystal Flash |
| **Head:** | Dark brown dubbing |

This is probably our favorite general-purpose midge pattern. The Krystal Flash gives the abdomen a segmented appearance and a slightly glossy sheen; at the same time, it produces a slim body that is characteristic of the naturals. This pattern penetrates the surface film easily but does not sink as quickly as the Brassie. The lighter weight allows the fly to drift more naturally, especially if it is affixed to the tippet with a loop knot. The fly can be fished deep as a trailer behind a heavier nymph, but it works particularly well fished near the surface, as a dropper behind a dry fly or small indicator. For this application, the fly can be tied on a dry-fly hook to further reduce the weight. Since Krystal Flash is slightly translucent, the thread foundation will influence the color of the body. Choose a thread that roughly approximates the color of the Krystal Flash. To tie a red-bodied version of this pattern, use red 8/0 thread and red Krystal Flash; omit the dubbing and form a simple thorax and head from the tying thread.

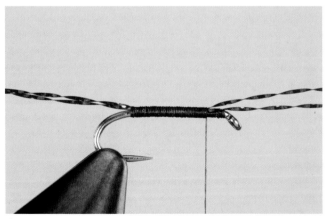

**1.** Mount the thread behind the hook eye, and wrap a smooth foundation to the rear of the shank. Mount 2 to 4 strands of Krystal Flash atop the shank. Flatten the thread by spinning the bobbin counterclockwise (when viewed from above), and secure the strands to the top of the shank by wrapping forward, forming a smooth underbody over the rear $\frac{4}{5}$ of the shank.

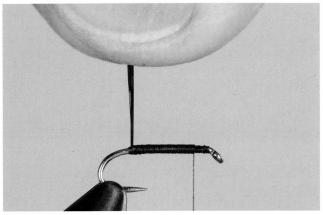

**2.** Clip the tag ends of the Krystal Flash. Pinch the strands in your right fingertips close to the hook shank. Draw your fingers upward to smooth, flatten, and consolidate the strands.

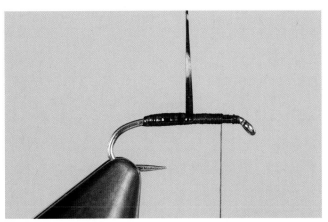

**3.** Wrap the strands forward in slightly overlapping turns, pausing as needed to pinch and draw the strands as described in step 2, which removes the twist from the Krystal Flash and keeps the wrapping strands flat.

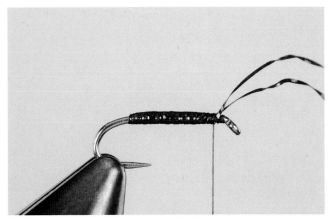

**4.** Continue wrapping until you reach the tying thread. Secure the strands.

**5.** Clip the excess Krystal Flash, and position the tying thread $1/5$ of a shank length behind the eye. With a sparsely dubbed thread, form a thorax about twice the diameter of the abdomen, stopping 2 to 3 thread wraps' distance behind the eye. Finish the head of the fly.

## Midge Pupa Patterns

**JUJUBEE MIDGE**

*Originator: Charlie Craven*

| | |
|---|---|
| **Hook:** | #16-20 scud |
| **Thread:** | White and black 8/0 |
| **Abdomen:** | 2 strands of olive, brown, red, or black Super Hair and 1 contrasting strand of black or white |
| **Wing case:** | White Fluoro Fiber |
| **Thorax:** | Black thread |
| **Wing buds:** | White Fluoro Fiber |

This Charlie Craven design is a model of fly-tying efficiency. Wrapping a contrasting strand of Super Hair along with the body-color strands essentially allows you to form and rib the abdomen in a single operation. The use of the wing-case material to form the wing buds provides a similar tying economy. Super Hair comes in a variety of colors, forms a slender abdomen, and is quite durable. This material is slightly translucent, and the color of the thread underbody will influence the appearance of the finished abdomen. Use a white thread underbody to preserve the original color of the Super Hair. If a darker abdomen is desired, use black thread for the underbody. This fly can be fished anywhere in the water column, but it works particularly well fished shallow, as a dropper tied 6 to 8 inches behind a dry fly or small indicator or used solo with the leader greased to within 6 to 8 inches of the fly.

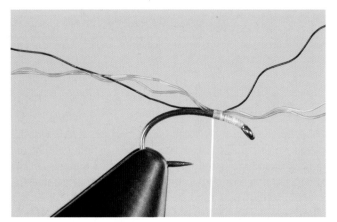

**1.** Mount the white thread behind the hook eye, and position the thread about ⅓ of a shank length behind the eye. Mount the 3 strands of Super Hair on top of the hook shank.

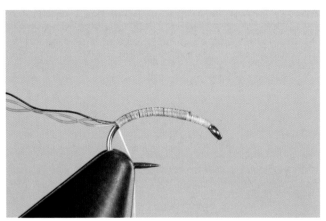

**2.** Trim the tags of Super Hair. Flatten the thread by spinning the bobbin counterclockwise (when viewed from above), and wrap rearward, securing the Super Hair strands to the top of the hook shank and forming a smooth thread underbody. Stop at the rear of the hook shank.

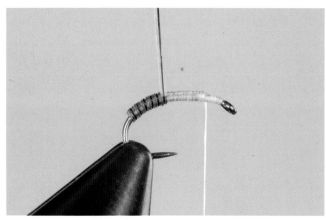

**3.** Wrap the thread forward to a point ¼ of a shank length behind the hook eye. Draw the strands of Super Hair upward to form a flat ribbon. Wrap the strands forward.

**4.** Secure the strands at the hanging thread. Tie off and cut the white thread. Mount the black thread behind the hook eye, and wrap it back to the front of the abdomen.

**5.** Mount about 12 to 20 strands of white Fluoro Fiber on top of the hook shank, and trim the excess. Wind the thread forward to form a smooth thorax slightly larger in diameter than the abdomen, stopping about 4 thread wraps' distance behind the hook eye.

**6.** Fold the Fluoro Fiber forward over the top of the hook shank, and secure it with 3 to 4 thread wraps.

**7.** Divide the strands of Fluoro Fiber into 2 equal bundles. Draw one bundle along the near side of the shank, and secure it with 2 thread wraps, as shown in this top view.

**8.** Draw the second bundle of fibers along the far side of the shank, and secure it. Wrap the thread rearward to the front of the wing case to lock the wing-bud fibers in place, as shown in this top view.

**9.** Finish the head of the fly. Draw both strands of Fluoro Fiber upward, and trim them just beyond the rear of the wing case.

## BEADHEAD THREAD MIDGE PUPA

| | |
|---|---|
| **Hook:** | #16-18 scud or standard nymph |
| **Head:** | $5/64$–$1/16$ inch black metal bead |
| **Thread:** | Black 8/0 |
| **Tail:** | Pearl Krystal Flash |
| **Rib:** | Fine red or copper wire |
| **Abdomen:** | Tying thread |
| **Wing case:** | Orange Antron yarn and Pearl Krystal Flash |
| **Thorax:** | Black tying thread |

This pattern is one of the many variations of the "thread midge," a generic design in which the body of the pupa is formed from the tying thread. Though this pattern can be fished anywhere in the water column, the extra weight from the metal bead makes it an ideal fly for fishing as a dropper in faster waters. We learned about using orange for the wing case from our angling friends in England, where midge (or "buzzer") fishing in stillwaters is a highly developed and widespread practice. The orange Antron does seem to work a bit better than the white we originally used, though gauging the effectiveness of a small change like this is admittedly difficult. On very small hooks, the ribbing wire can appear disproportionately thick and overwhelm the body. You can decrease the wire diameter by repeatedly drawing a length of wire gently through your pinched fingertips to stretch and thin the strand.

**1.** Slide the bead onto the hook, position it behind the hook eye, and place the hook in the vise. Mount the thread behind the bead, and secure 2 strands of Krystal Flash and the ribbing wire to the top of the hook shank. The tag ends of the Krystal Flash strands should extend about 1 inch beyond the hook bend and hook eye.

**2.** Wrap to the rear of the shank, securing the Krystal Flash strands and wire atop the hook shank.

**3.** Trim the wire tag. Wrap the thread forward, forming a slightly tapered body over the rear 2/3 of the hook shank. Position the thread about 1/3 of a shank length behind the hook eye.

**4.** Spiral the ribbing wire forward in 6 to 7 turns to the tying thread. Secure it with 3 to 4 thread wraps, and trim the excess wire. Position the thread directly behind the bead.

**5.** Separate the strands of Krystal Flash, and pull one strand to each side of the hook shank, as shown in this top view. Position the Antron yarn fibers on top of the hook shank so that the forward-facing fibers extend about 1½ inches beyond the hook eye. Mount the Antron atop the shank, between the Krystal Flash strands, directly behind the bead. Trim the rear tag of yarn. Wrap the thread rearward, securing the Antron to the top of the shank and forming a thorax slightly larger in diameter than the abdomen. Finish the thorax at the rib tie-off position.

**6.** Draw the bundle of Antron fibers rearward over the top of the thorax, and secure it with 3 tight thread wraps. Draw the Krystal Flash rearward, placing one strand along each side of the Antron wing case. Secure the strands.

**7.** Tie off the thread over the wraps used to secure the Antron and Krystal Flash, and clip the thread. Trim the rearmost tags of Krystal Flash to make a tail about ½ the width of the hook gap. Trim the Antron and Krystal Flash forming the wing case to leave a short tag, as shown. Then coat the tie-off wraps with head cement.

# Midge Emerger Patterns

## PUFF MIDGE

| Hook: | #16-22 standard dry-fly |
|---|---|
| Thread: | Black, brown, or green 8/0 |
| Tail: | Medium dun CDC feather barbs |
| Abdomen: | Tying thread |
| Wing: | Dark dun or black CDC feather barbs |
| Thorax: | Fine black or dark brown dubbing |

The method used here for mounting the CDC barbs, which produces no underbody bulk behind the wing-mounting point, along with the thread abdomen and sparsely dubbed thorax, gives this fly a very slender body like that of the natural. Choose a soft, fine-fibered dubbing such as mole or beaver (with the guard hairs removed) or a superfine poly dubbing. Moisten your fingertips and twist the dubbing tightly to keep the thorax smooth and trim rather than rough and shaggy. This is a highly productive pattern when trout are feeding on midge pupae in or just below the surface film. Under normal conditions, we use a dark dun wing, but under low- or flat-light conditions, the black CDC is noticeably more visible to the angler.

Like the wings on most CDC patterns, the puff on this fly will probably become matted after a couple of fish, in which case you're best off tying on a fresh fly. Rinse the wing of the used fly, and blow away the excess moisture. After it's thoroughly dry, it can be fished again. The procedure for dressing this fly is virtually identical to that used to tie the BWO Puff Fly, pp. 8–10, so the steps here are abbreviated.

**1.** Mount the thread behind the hook eye, and wrap to the rear of the shank. Mount 10 to 12 CDC barbs atop the shank. Wrap the tying thread forward to form a slightly tapered abdomen over the rear half of the shank. Position the thread at the front of the abdomen.

**2.** Trim the tail to be about one hook gap in length. Prepare a bundle of CDC barbs as described for the BWO Puff Fly, steps 1–3, p. 9. Mount the barbs and post them upward as shown in steps 4–5.

**3.** Use a very sparsely dubbed thread to form the thorax.

**4.** Finish the fly behind the hook eye. Draw the wing barbs upward, and trim them to a height of about one shank length.

## MIDGE EMERGER

| | |
|---|---|
| **Hook:** | #16-20 TMC 206BL |
| **Thread:** | Black 8/0 |
| **Tail:** | White Antron fibers |
| **Abdomen:** | Tying thread |
| **Thorax:** | Fine black dubbing |
| **Wing post:** | White poly yarn |
| **Hackle:** | Dark dun dry-fly |

This midge version of the Klinkhamer design has some distinct advantages. Because the body is suspended underwater beneath the hackle, it rather nicely imitates a midge pupa tucked up against the surface film, and it's been a productive pattern for us when trout are feeding at or just below the surface. The white wing post makes this fly easier to see on the water than most emerging midge patterns, and it floats well enough to support an unweighted pupa pattern fished as a dropper. Hackling this fly in the smallest sizes can be a bit tricky; since relatively little poly yarn is used to form the wing post on a small hook, the post can bend rather easily under hackle-wrapping pressure, and the wraps of hackle can slip off. For the best results, take care to form a tight thread foundation around the base of the post to make it rigid, and use good-quality, genetic saddle hackle, which has a fine, flexible stem that requires little pressure to wrap. Since the dressing of this pattern is almost identical to procedure for the Klinkhamer Female Trico on p. 49, the steps here are abbreviated.

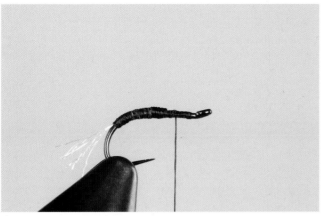

**1.** Mount the tying thread behind the hook eye, and wrap to the rear of the shank. Mount a sparse bundle of Antron fibers atop the shank. Trim the excess, and clip the rear fibers to make a trailing shuck about one hook gap in length. Wrap the tying thread forward to form a slightly tapered abdomen over the rear $^2/_3$ of the shank. Position the thread in front of the abdomen.

**2.** Mount the wing post and hackle using the procedure shown for the Parachute BWO, steps 1–4, pp. 15–16.

**3.** Dub a thorax slightly larger in diameter than the abdomen, ending with the tying thread behind the hook eye.

**4.** Wrap and tie off the hackle using the procedure shown for the Parachute BWO, steps 7–10, pp. 16–17. When the hackle tip is secured, trim the excess and finish the head. Draw the wing post upward, and trim it to about one shank length in height.

## Midge Adult Patterns

**PARACHUTE MIDGE**

| | |
|---|---|
| **Hook:** | #16-22 standard dry-fly |
| **Thread:** | Black 8/0 |
| **Tail:** | Dark dun Microfibetts |
| **Body:** | Fine black dubbing |
| **Wing post:** | White poly yarn |
| **Hackle:** | Dun dry-fly |

When the trout are looking up and taking individual midge adults from the surface, this is the first fly we tie on. This pattern is virtually identical to the one we use for imitating male Trico spinners, though in this case, we use a straight, rather than triple, tail. The white wing post is fairly easy to see on placid currents or in good light, but under less than optimum conditions, it can be a bit difficult to pick out on the water. We have dressed this fly with an orange wing post, which is more visible, but we've become convinced that on occasion, a trout will reject the fly apparently because of the bright spot of color. But on the whole, you may find that the ability to see the fly, manage its drift, and spot the strike is worth a few rejections. As noted above for the Midge Emerger, parachute hackling the smallest hook sizes can be made easier by forming a tight thread foundation around the base of the wing post and using supple, fine-stemmed hackle. The procedure for tying this fly is identical to that shown for the Parachute BWO, pp. 15–17, though the body on this pattern should be kept quite slender, and when the fly is finished, the wing post should be trimmed to about ½ a shank length in height.

## SNOWSHOE MIDGE CLUSTER

| Hook: | #16-20 standard dry-fly |
|---|---|
| Thread: | Black 8/0 |
| Hackle: | Dark dun dry-fly |
| Body: | Peacock herl |
| Wing: | Black or dark dun snowshoe hare's foot hair |

Like the popular Griffith's Gnat, this pattern is designed to suggest a swarm of midges congregated on the surface of the water. While the Griffith's Gnat is certainly a worthy pattern, we prefer the advantages that this fly offers. The snowshoe hare wing makes this pattern significantly easier to see, and it fishes better on choppy water. It floats well enough to support a heavier subsurface dropper, like the Beadhead Thread Midge Pupa, and it can double as a small caddis imitation. This is our top choice for when midges are clustering on the surface and small rafts of them break free to drift downcurrent, or for fishing riffles when midge are active. Avoid the temptation to overdress this pattern, especially in the smallest sizes. Though snowshoe hare's foot hair doesn't flare like deer hair, it does fluff out a good bit when tied down, and it's easy to make an overly heavy wing. It doesn't take much material to float a small dry-fly hook. Choose peacock herl that is proportional to the hook size; strands taken from nearer the base of an eyed feather or from a peacock sword are usually narrower and better suited to small hooks. This pattern is tied using the steps for the Black Snowshoe Caddis, p. 148.

## SPENT MIDGE

| Hook: | #16-22 standard nymph or dry-fly |
|---|---|
| Thread: | Black 8/0 |
| Rib: | Fine copper wire |
| Body: | Tying thread |
| Wings: | Tan CDC and white Antron yarn |

Because female midges typically crawl underwater to lay their eggs, many of them end up in the subsurface drift. Whether trout key in on these drowned adults is difficult to say for certain; it's impossible to observe the fish and these tiny insects closely enough. But this fly, tied to imitate a spent adult, is a productive one, particularly when fish are ignoring patterns that represent other stages of the midge life cycle. We actually tie this fly in two versions. The first is sparsely dressed on a nymph hook for fishing subsurface, where the mobility of the CDC suggests the limp, flexible wings and legs of a midge tumbling in the current. The second is tied with a denser wing on a dry-fly hook, since some spent midges can be observed drifting on the surface film. In either version, however, this pattern can be devilishly difficult to see and is best fished as a dropper behind a small indicator or dry fly.

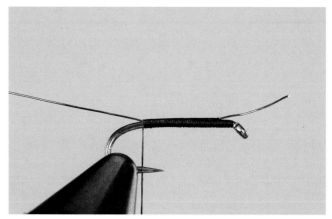

**1.** Mount the thread behind the hook eye, and position it about 5 to 7 thread wraps' distance behind the eye. Mount the ribbing wire atop the shank, and then bind it down in close, tight wraps to the rear of the hook shank.

**2.** Trim the front tag of the wire. Wrap the thread forward to form a smooth body over the rear ⁴/₅ of the hook shank.

**3.** Spiral the ribbing wire forward to the tying thread. Tie off and trim the wire. Position the thread halfway between the front of the body and the hook eye.

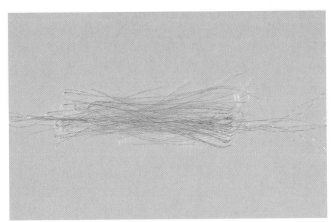

**4.** Strip the desired quantity of barbs from a CDC feather as described for the BWO Puff Fly, steps 1–3, p. 9. Divide the barbs into 2 equal bundles, and then rotate one bundle 180 degrees so that the tips are aligned with the butts of the other bundle. Then place 4 to 6 Antron fibers on top of the CDC fibers.

**5.** Gather the CDC barbs and Antron fibers into a bundle. Secure the midpoint of the bundle to the top of the hook shank with 2 thread wraps.

**6.** Rotate the bundle so that it is perpendicular to the hook shank; secure it using crisscross wraps, as shown in this top view. Position the thread directly ahead of the wing.

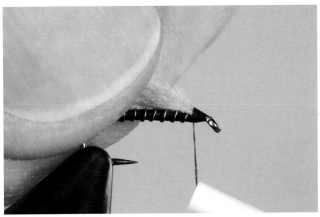

**7.** Fold both wings rearward and slightly upward. Take 2 to 3 tight thread wraps over the base of the fold to hold the wings in position.

**8.** Preen the wings outward so that they flare out about 45 degrees from the body, as shown in this top view. Take a few more thread wraps over the base of the folded wing fibers to lock them in place, and then finish the head.

**9.** Gather the wings together, and trim them about even with the hook bend.

# Chapter Five

# *Terrestrials*

## MAJOR TERRESTRIALS

## Ants

Biologist Edward O. Wilson has estimated that the total weight of all the ants on earth is greater than the total weight of all the humans. That's a lot of ants. So it's not surprising that some of them find their way into trout streams. For fly anglers, the most important ants are found in large colonies, typically those that live in rotting wood, below ground, or in anthills. Like honeybees, ants are social insects, and a colony consists of an egg-laying queen, winged males that will mate with new queens, and workers, sterile females that make up most of the colony. New queens emerge with wings, and at this time the winged males leave the colony, fly into the air, and form a mating swarm. The new queen flies into the swarm to mate, after which she lands on the ground, loses her wings, returns to the colony or starts a new one, and begins laying eggs.

Even though the wingless workers make up most of a colony, they do not routinely end up in the water, though it certainly happens on occasion. But the winged males that form the mating swarms often land or fall on the water, and trout readily feed on them. An ant "hatch" like this is, in some respects, among both the most and least predictable events on a trout stream: it will almost certainly occur at some point, but you never know exactly when. Ants tend to form new colonies at the edges of the season, so mating flights are most apt to take place in the spring and fall, but it's not uncommon to find them during the summer months as well. Moreover, ants are found throughout the West, and an angler who doesn't carry a few ant patterns at all times can miss out on some excellent fishing. When ants are on the water—and it doesn't take many of them—trout feed selectively on these insects. Some anglers have speculated that the fish particularly relish the taste of ants.

| | |
|---|---|
| **Other common names:** | Carpenter ants, field ants, mound-building ants, red/black ants, black ants, red ants |
| **Order:** | Hymenoptera |
| **Family:** | Formicidae |
| **Mating flight:** | April to September |
| **Body length:** | ¼–¾ inch (6–20 mm); carpenter ant workers and winged males are about ½ inch long; red/black, red, and small black ants are about ¼ inch long |
| **Hook sizes:** | #12-18 |

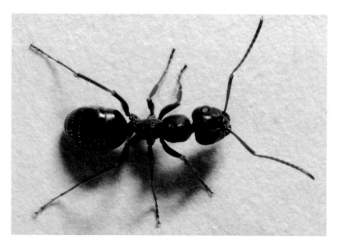

**Above:** *Black carpenter ant worker.*

**Above:** *Red/black ant worker.*

**Left:** *Winged male black ant.*

**Above:** *Winged male black ant (underside).*

**Left:** *Winged male red/black ant.*

**Above:** *Winged male red/black ant (underside).*

## Important Fishing Stages

**Wingless worker** Unless they are dislodged from bankside vegetation or blown into the stream, neither of which seems to be terribly common, worker ants are seldom found on the water. That having been said, it's difficult to explain why wingless ant patterns work so well. Through the years, we've seen many fish taken on floating and sunken wingless ants. Perhaps trout do indeed love the taste of them, maybe they have a generalized buggy appearance that interests the fish, or possibly there is some other reason. But wingless ant patterns catch trout. The most reliable way to fish one of these flies is to trail it as a dropper behind a small indicator or dry fly, since even the floating patterns are hard to spot on the water. Wingless ant patterns are generally most effective during the warmer part of the day in the spring, summer, and fall when ants are active and no other insects are present. But fishing a sunken or floating ant when mayflies or caddis are present can bring surprising results. Ken Miyata, a fisherman whose name is well known to anglers who came of age before or during the 1980s, extolled the virtues of fishing ant patterns during hatches of other insects.

**Winged male** Unlike mating swarms of caddis and mayflies, which collect over the water, swarms of flying ants congregate high enough above the ground that they are almost invisible. Such swarms are often extremely localized and short in duration as well. The most reliable indicator of flying ants is the presence of winged males on the ground along a stream, though these can be easily missed if you're not looking for them. The problem of recognizing an ant hatch is compounded by the reality that winged males won't necessarily land on the water directly in front of you. They may well come drifting downstream from above. Ants often sink, and those that float ride flush in the film and can be almost impossible to see on the water unless they are close to you. So should you see trout rising to some unseen insect during the warmer part of the day in the spring or fall or midmorning in the summer, inspect the water closely for drifting ants. If you discover them, match your fly to the size and color. The type of ant you might find depends to some extent on the area you're fishing. The larger black carpenter ants and red/black mound-building ants are often found along wooded streams; red/black and red ants are common in drier habitats.

## Wingless Ant Patterns

**THREAD ANT**

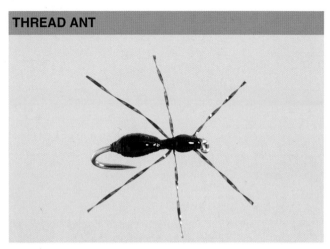

| Hook: | #12-18 standard nymph |
| Thread: | Black or red and black 6/0 or 8/0 |
| Abdomen: | Black or red thread |
| Thorax: | Black or red thread |
| Legs: | Black or rusty brown Krystal Flash |

Requiring only two materials, one of which is the tying thread itself, this pattern is among the simplest in fly tying. Black is the most common color for the fly, but a bicolor pattern with black abdomen and red thorax can be fashioned simply by changing thread colors during tying. When the fly is complete, coat the abdomen and thorax with Sally Hansen Hard as Nails clear nail polish or a thick head cement to improve durability and give the body a shiny appearance like that of the natural. Designed to be fished subsurface, this pattern sinks quickly and is best fished as a dropper behind a dry fly or indicator.

**1.** Mount the thread behind the hook eye, and wrap to the rear of the hook shank in close, tight turns. Layer the thread to form a tapered abdomen over the rear half of the hook shank.

**2.** Position the thread ⅓ of a shank length behind the hook eye. Form a small bump with 3 or 4 wraps of thread. Mount a strand of Krystal Flash on the near side of the hook shank in front of the thread bump, as shown in this top view.

**3.** Fold the forward-facing strand of Krystal Flash over the hook shank to the far side, and secure it with thread by wrapping rearward over the thread bump. The back legs should now angle rearward, as shown in this top view.

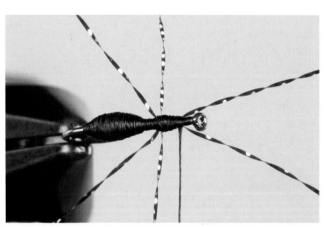

**4.** Position the thread directly in front of the thread bump securing the rear legs. Mount a strand of Krystal Flash on each side of the hook shank. Hold the front tag ends of Krystal Flash against the sides of the hook shank, and wrap the thread forward, stopping about 6 thread wraps' distance behind the hook eye.

**5.** Use wraps of thread to form a tapered thorax/head, ending behind the hook eye.

**6.** Tie off and cut the thread. Trim the legs to about the length of the hook shank. Then coat the thread body with Sally Hansen Hard as Nails nail polish or a thick head cement.

## FUR ANT

| | |
|---|---|
| **Hook:** | #12-18 standard dry-fly |
| **Thread:** | Black 6/0 or 8/0 |
| **Abdomen:** | Fine black dubbing |
| **Hackle:** | Black or dark dun dry-fly |
| **Thorax:** | Fine black dubbing |

This simple pattern is more versatile than the Thread Ant, as it can be fished either wet or dry. Treated with floatant, it rides flush against the surface film; left untreated, the pattern sinks slowly, as the naturals do. Black is our first color choice, but a dark brown or cinnamon body with brown hackle is also effective. The natural insect has a smooth, well-delineated body profile; twist the body dubbing tightly around the thread to reproduce this clean silhouette. The dressing is designed to be sparse; only a few wraps of hackle are needed.

**1.** Mount the thread behind the hook eye, and wrap it in close, tight turns to the rear of the shank. Dub back and forth in layers to form a tapered abdomen over the rear half of the hook shank. Position the thread directly in front of the abdomen.

**2.** Prepare and mount the hackle as shown in the instructions for the Little Green Drake, steps 4–6, p. 42. Position the thread halfway between the front of the abdomen and the hook eye.

**3.** Take 4 to 6 wraps of hackle, and tie off the feather as shown in steps 7–10, pp. 42–43.

**4.** Trim away the hackle barbs that project beneath the hook shank, as shown in this front view.

**5.** Use your left fingers to gently draw back the hackle. Dub the thorax/head portion of the fly, finishing 3 to 4 thread wraps' distance behind the eye. Finish the head.

# Winged Ant Patterns

## DELTA ANT

| Hook: | #12-18 standard dry-fly |
|---|---|
| **Thread:** | Black or dark brown 6/0 or 8/0 |
| **Body:** | Fine dubbing in black, cinnamon, or rusty red |
| **Wings:** | Tan Antron yarn |
| **Hackle:** | Black, dark dun, or brown dry-fly |

There are a great many flying-ant patterns, some of them quite elegant and meticulous. But in the end, we prefer this more workmanlike design—it is easy and relatively quick to tie, durable, and catches trout well. Though the hackle on the underside of the hook shank is trimmed to make the fly ride flush against the surface, the white wings make larger sizes of this fly reasonably easy to spot on the water. In small sizes, however, the fly can be difficult to see, and we often fish it as a dropper behind a larger ant or other dry fly. We prefer fine poly dubbing for the body because it resists water absorption. Even so, the pattern lacks any truly buoyant materials, and the entire fly should be treated with floatant.

**1.** Mount the thread behind the hook eye, and wrap a thread underbody to the rear of the shank. Work back and forth with a dubbed thread to build a tapered abdomen over the rear half of the shank. Position the thread midway between the front of the abdomen and the hook eye.

**2.** Cut a strand of Antron yarn about 2 inches long. If necessary, split the strand lengthwise for wings on smaller hooks. Center the strand over the tying thread, and secure it to the near side of the hook shank.

**3.** Fold the forward-facing strand to the rear, and secure it against the far side of the shank by taking tight thread wraps over the fold in the yarn, as shown in this top view.

**4.** Prepare and mount a hackle as shown in the instructions for the Little Green Drake, steps 4–6, p. 42.

**5.** Take 3 to 4 wraps of hackle, and tie off the feather as shown in steps 7–10, pp. 42–43.

**6.** Dub the thorax/head portion to within 3 thread wraps' distance of the hook eye.

**7.** Finish the head of the fly, and clip the thread. Draw both wings rearward at once, and trim them even with the hook bend, as shown here from the top.

**8.** Trim the hackle beneath the shank flush with the bottom of the fly, as shown in this front view.

## PARACHUTE ANT

| | |
|---|---|
| **Hook:** | #12-18 standard dry-fly |
| **Thread:** | Black or dark brown 6/0 or 8/0 |
| **Body:** | Fine dubbing in black, cinnamon, or rusty red |
| **Wing post:** | White or red poly yarn |
| **Hackle:** | Light or medium dun dry-fly, one size larger than normally used for the hook size |

On this ant pattern, as on some mayfly spinner patterns, parachute hackle nicely imitates delicate, translucent wings pressed flat against the surface film. The parachute style has other advantages as well. The wing post makes the fly relatively easy to see on the water, and like most parachute patterns, this one floats well, making it a good choice for choppy or riffled water. On smaller hook sizes, however, leave the wing post taller than noted in step 6 for better visibility. You can trim it down if necessary on the stream. On smooth, flat water, we prefer a white wing post, believing that brighter colors occasionally cause trout to reject the fly.

**1.** Mount the thread behind the hook eye, and wrap a thread foundation to the rear of the shank. Work back and forth with a dubbed thread to build up a tapered abdomen in layers over the rear half of the shank. Position the tying thread midway between the front of the abdomen and the hook eye.

**2.** Using the instructions for the Parachute BWO, steps 1–2, p. 15, mount the wing post.

**3.** Prepare and mount a parachute hackle as shown in steps 3–4, p. 16, but do not prop the feather in front of the wing post.

**4.** Dub the thorax in front of the wing post. Position the thread directly in front of the thorax.

**5.** Wrap and tie off the hackle as shown in steps 7–10, p. 16–17.

**6.** Clip the feather tip, and finish the fly. Trim the wing post to about ½ the hook gap in height. Clip away the hackle barbs at the front of the wing post to make a gap of about 90 degrees centered over the hook shank, as shown in this top view.

# Beetles

With more than 300,000 species of beetles worldwide, the order Coleoptera is estimated to comprise a quarter of all the known insects, a fact that allegedly moved evolutionary biologist J. B. S. Haldane to remark, "It would appear that God has an inordinate fondness for beetles." There are more than 30,000 species in North America alone, some aquatic and some terrestrial. But only the adult terrestrial beetles that move about, fly, and periodically end up on the water are important to stream anglers. You won't normally see large numbers of these insects, but during the season, enough of them end up in the water that the trout recognize them as food and take them readily. Under most circumstances, beetles are employed as searching patterns, used to fish the water rather than cast to rising trout. In high-elevation streams, beetles, along with moths, are generally the largest terrestrial insects available to trout, and beetle patterns can be highly effective on these waters. Though there is no true "hatch" of these insects, we consider adult beetles important because they are active along virtually all trout streams in the spring, summer, and fall; because they wind up in the water sufficiently often to keep the trout interested; and because beetle patterns are responsible for a great many fish every year.

| | |
|---|---|
| **Order:** | Coleoptera |
| **Availability:** | Spring through fall |
| **Body length:** | $\frac{1}{16}$–$1\frac{1}{2}$ inches (1–40 mm) |
| **Hook sizes:** | #8-18 |

**Above:** *Beetle adult: beetles can differ a good deal in size and body shape. The insect shown here is fairly large, about a size 10, but the profile is characteristic of many beetles, large and small.*

**Top right:** *Beetle adult (underside).*

**Right:** *Beetle adult: body colors vary widely; black, brown, and green are common, but some insects are brightly colored or patterned.*

# Important Fishing Stages

**Adult** Only the adult stage of the beetle life cycle is significant to anglers, and even then its importance varies. Though beetles are common along trout streams, you seldom see large numbers of them unless there is a local infestation and the adults are migrating to colonize new areas. These outbreaks typically occur in wooded landscapes, and to fishermen they can be a blessing or a curse. Trout may become highly selective at these times, and the wide range of sizes, shapes, and colors of beetles can pose a problem, since most anglers don't carry a large variety of beetle patterns. But if your fly boxes are well stocked, or you happen to get lucky and have the right pattern, you can experience some fine fishing in beetle-infested areas. More often, though, anglers use a beetle pattern to prospect for trout, looking for the opportunistic feeders, and as a rule, these fish are not often selective. Carrying a black beetle pattern in a few different sizes usually suffices. It's always wise, however, to have smaller beetles in size 16 or 18 as well; trout that reject hatch-matching patterns during an emergence will take small beetle patterns, as they do ant patterns, surprisingly often. In fact, a useful technique on heavily fished waters is to use a small beetle pattern, by itself or with a nymph or emerger imitation, during hatches of mayflies or caddis.

# Beetle Patterns

**FOAM BEETLE**

| | |
|---|---|
| **Hook:** | #8-18 standard dry-fly |
| **Thread:** | Black 6/0 or 8/0 |
| **Overbody:** | Black Larva Lace Dry Fly Foam |
| **Underbody:** | Peacock herl or Antron dubbing to match natural |
| **Legs:** | Black round rubber or silicone |
| **Indicator (optional):** | Orange poly yarn |

The very first closed-cell foam trout flies we saw, decades ago, were beetle patterns, and the suitability of the material for the purpose was instantly apparent. Foam ties easily; it forms a smooth, shell-like, realistic overbody; and it requires no floatant. A larger foam beetle pattern has enough buoyancy to support a nymph dropper. There are a great many foam-bodied beetle patterns these days, some of them quite elaborate, but this simple design has served us well. The shape of the overbody can be varied: a narrow strip of foam, as shown in the following sequence, produces an elongated body; a wider strip makes a thicker, more rounded one. The leg material should be supple enough to wiggle a bit as the fly drifts; choose rubber legs that are thin and flexible, or use silicone leg material, which is limper than rubber. By changing the color of the foam and the underbody on this pattern, you can match locally prevailing species of beetles or those in an infested area. The larger sizes of this fly are easy to see on the water without the indicator yarn. But the pattern does float low, and the visibility of smaller flies can be improved by adding the indicator. Instead of the Larva Lace, a strip of a similar $\frac{1}{16}$-inch (2 mm) closed-cell foam can be substituted. For information on types of foam suitable for this pattern, see p. 79.

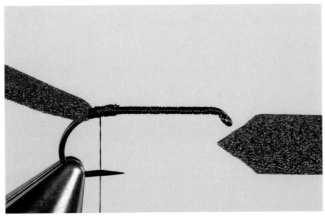

**1.** Mount the thread behind the hook eye, and wrap a thread foundation to the rear of the shank. Cut a 2-inch strip of foam that is one hook gap in width, and trim one end to a point, as shown. Mount the point of the foam atop the shank, and bind down the excess.

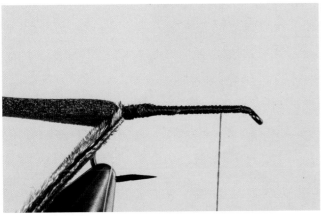

**2.** Position the thread at the rearmost thread wrap securing the foam. Mount 3 to 5 peacock herls and a 6-inch length of scrap tying thread. Trim the excess, and position the tying thread ¼ of a shank length behind the hook eye.

**3.** Twist the herls and thread together to make a chenille-like strand. Wrap the herl forward to the tying thread, pausing as needed to retwist the strand. Secure the herl at the hanging thread.

**4.** Clip the excess herl strand. Fold the foam strip forward over the top of the shank, and secure it with 4 to 5 thread wraps.

**5.** Draw the tag of foam forward, and trim it even with the hook eye.

**6.** Use scissors to round the corners of the head. Cut 2 lengths of leg material twice as long as the hook shank. Mount them to the sides of the shank as described for the Green Drake Rubber Legs, steps 6–9, pp. 58–59, and shown in this top view.

**7.** To tie the optional indicator, center a 1-inch length of yarn over the leg-mounting wraps, and bind it to the top of the hook shank.

**8.** Pull both ends of the yarn upward, and trim them to about ⅓ the hook gap in height. Finish the fly over the yarn-mounting wraps.

## CROWE BEETLE

*Originator: John Crowe*

| | |
|---|---|
| **Hook:** | #12-18 standard or 1XL dry-fly |
| **Thread:** | Black 6/0 or 8/0 |
| **Body:** | Black deer hair |
| **Head:** | Black deer hair |
| **Indicator (optional):** | Orange poly yarn |

The Crowe Beetle is decidedly old-school, and anglers who grew up in the age of foam may never have heard of this pattern. But the dyed deer hair used for the body floats well and has a glossy sheen like the shell of a natural beetle, and in ways that are difficult to explain, this pattern seems to sit on the water more realistically than foam types. From the tying standpoint, we particularly favor this fly in the smaller sizes, since proportioning a foam strip for a small hook can be tedious. The Crowe Beetle is our first choice for fishing a terrestrial during a mayfly or caddis emergence. The deer hair does not make an exceptionally durable body, but it's one of those patterns that seem to work better when they're a little roughed up and scraggly. And it's easy to tie; in the smallest sizes, you can omit the legs. The fly floats flush against the surface, and if you have difficulty spotting it on the water, add the orange indicator yarn as shown for the Foam Beetle, steps 7–8 (at left).

**1.** Mount the thread behind the hook eye, and wrap a tight thread foundation to the rear of the shank. Clip and clean a bundle of deer hair as described for the Sparkle Dun, steps 1–2, p. 27. There's no need to stack the hair, but trim the hair butts so that they are even.

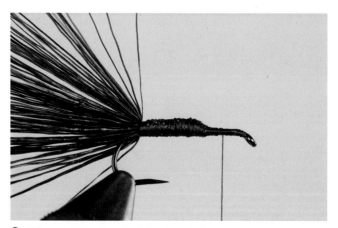

**2.** Mount the butt end of the deer-hair bundle atop the shank, and bind down the excess. Position the tying thread ¼ of a shank length behind the hook eye.

**3.** Draw the bundle of hair over the top of the shank to make a smooth, shell-like overbody. Secure the hair atop the shank with 4 to 5 tight wraps. Don't trim the excess.

**4.** Separate 3 or 4 strands of hair from the far edge of the clump extending over the hook eye, and fold them rearward along the far side of the hook shank. Then separate 3 or 4 strands from the near side of the clump, and fold them rearward along the near side of the shank. Take 3 to 4 thread wraps at the base of the folded strands to stand the hairs perpendicular to the hook shank on either side, as shown in this top view.

**5.** Lift the tuft of hair over the hook, and take 2 thread wraps around the shank. Then finish the fly around the shank behind the eye.

**6.** Pull the tuft of deer hair extending over the hook eye gently forward, and trim it, following the angle of the hook eye as a guide.

**7.** Trim the legs to about one hook gap in length, as shown in this top view.

**8.** To crinkle the legs for a more realistic appearance, use your thumb and forefinger to pinch gently inward on the tips of the legs, as shown here from the top.

# Grasshoppers

The grasshoppers most commonly found along Western streams belong to the family Acrididae and are commonly called short-horned grasshoppers because of their short antennae. They live in grasslands, rangelands, and open woods, where they feed on grasses and, much to the consternation of farmers and ranchers, crop plants as well. The nymphs emerge from the ground in spring and look like miniature adults, but at this point they lack wings and sex organs. When these grasshoppers reach maturity in mid to late summer, they are able to fly and mate. The mature adults feed actively and often hop or fly to new areas in search of food, and these flying or migrating grasshoppers are usually the ones that end up in the water, much to the delight of the trout. The grasshoppers—and the grasshopper fishing—remain lively until the cold weather of fall puts an end to both. Arguably, more large trout are caught on hopper patterns than on any other dry fly, partly because adult grasshoppers are big and meaty, and partly because the hopper season is a relatively long one that can hold up well into the fall during warmer years. Grasshoppers are unquestionably the best terrestrial super hatch, and many anglers consider hopper fishing the most exciting and satisfying dry-fly fishing of the season. The fly patterns are generally large and visible, and the strikes spectacular.

**Other common names:** Hopper, Lesser Migratory Locust
**Order:** Orthoptera
**Family:** Acrididae
**Emergence:** July to September
**Body length:** ¾–2½ inches (18–64 mm)
**Hook sizes:** #6-12, 2XL–3XL

**Left:** *Grasshopper adult with yellow body.*

**Above:** *Grasshopper adult with yellow body (underside).*

**Left:** *Grasshopper adult with tan body.*

**Above:** *Grasshopper adult with tan body (underside).*

**Left:** *Grasshopper adult with gray body.*

**Above:** *Grasshopper adult with gray body (underside).*

# Important Fishing Stages

**Adult** Immature grasshoppers are flightless and seldom found on the water. But from midsummer on, when they reach maturity and begin flying, grasshoppers frequently end up in the water. The summer winds are most definitely a contributing factor, blowing airborne hoppers off course or knocking them into the stream. Grasshoppers ordinarily don't start flying until the air temperature reaches the low 70s, around midmorning in the summer months, and this is generally the best time to start using a hopper pattern. Even so, trout respond to grasshoppers much like they do Salmonflies and Golden Stoneflies; once the hatch is under way, the fish become accustomed to seeing these big bugs on the water, and you can often fish a hopper pattern productively any time during the daylight hours. Casting to cutbanks or runs close to the shore is the classic hopper-fishing scenario, but it's worth remembering that flying grasshoppers can end up anywhere on the surface, so it pays to cover any likely looking trout water. It's not at all uncommon for trout holding in deeper pools and midriver runs to rise to a hopper pattern. Fishing a dropper pattern behind a grasshopper imitation is a highly effective technique early in the morning and for prospecting riffles any time of day. It's a good idea to match a hopper pattern to the approximate size and color of the naturals found along the stream, since grasshoppers vary some in appearance, and a close imitation is often what it takes to catch wary trout. It's common practice throughout the West to splat big, chunky hopper patterns down on the surface, and this technique can produce some dramatic strikes, particularly later in the year. But an underused approach is to fish relatively small grasshopper imitations—about 1 to 1 ¼ inches long—especially in the first half of the season. It may only be that the trout on hard-fished waters are less accustomed to seeing flies of this size and thus are less suspicious. Whatever the reason, smaller hopper imitations have proven quite productive for us.

# Grasshopper Patterns

**PARACHUTE HOPPER**

*Originator: Ed Schroeder*

| | |
|---|---|
| **Hook:** | #8-14 2XL dry-fly |
| **Thread:** | Brown 6/0 |
| **Wing post:** | White calf tail |
| **Hackle:** | Grizzly dry-fly |
| **Abdomen:** | Amber poly dubbing |
| **Wing:** | Mottled turkey quill section |
| **Rear legs:** | Preknotted pheasant tail feather |
| **Front legs (optional):** | Small brown or tan round rubber or silicone leg material |
| **Thorax:** | Amber poly dubbing |

This parachute design floats well, and the white wing post is quite visible, making the pattern suitable for fishing choppy or broken water. But this flush-floating fly has a highly realistic silhouette when viewed from underneath, and it's particularly useful on spring creeks and on the flatter, smoother sections of other streams, where the trout get a long, close look at it. The rubber legs are optional; we use them on the larger hook sizes for a more accurate imitation of the natural, since the front legs on a big hopper are conspicuous. This fly is not particularly difficult to tie, but it does have a lot of components, and it may take a few attempts to get the proportions correct. It's more efficient to cut and cement all the wing segments at once, prior to tying, than to prepare each one separately as you dress individual flies.

**1.** Mount the thread behind the hook eye. Wrap a thread foundation to the rear of the shank, and return the thread to a point about ¼ of a shank length behind the eye. Clean and align a bundle of calf tail as described for the Parachute Adams, step 1, p. 97. Mount the bundle so that the hair tips extending forward of the mounting wraps are ⅔ the length of the hook shank.

**2.** Clip and bind down the hair butts. Lift the wing fibers vertically. Take several tight thread wraps against the front base of the bundle, building a bump of thread that will keep the wing fibers upright. Build a foundation for the hackle with 10 to 12 thread wraps, wrapping up the wing post; then take another 10 to 12 wraps down to the base of the post. Position the thread behind the wing post.

**3.** Prepare and mount the hackle as described for the Parachute BWO, steps 3–4, p. 16.

**4.** Wrap the thread to the rear of the shank. Dub a thick, slightly tapered abdomen up to the rear of the wing post. The final wraps of dubbing should abut the wing post without a gap.

**5.** Cut a section of mottled turkey wing quill about one hook gap in width. Coat the strip on both sides with a flexible fly-tying cement, and let it dry. Round the corners of the thick end of the feather strip as shown. Trim the thin end of the feather strip to make a wing that extends from the base of the wing post to just beyond the hook bend.

**6.** Center the wing strip along the top of the shank, abutting the wing post, and fold it down the sides of the body. Flatten the thread by spinning the bobbin counterclockwise (when viewed from above). Secure the wing strip atop the shank and, at the same time, form a smooth foundation, about 4 to 6 thread wraps' wide, behind the wing post. The legs will be mounted against these thread wraps, and having a reasonably even, uniform foundation will simplify the mounting and give the best results. Position the thread at the rearmost thread wrap securing the wing.

**7.** Mount a preknotted pheasant tail leg on the near side of the shank. The leg should angle slightly upward, as shown, and the knot in the pheasant barbs should be aligned with the rearmost point of the abdomen.

**8.** Mount the second leg on the far side of the shank so that it matches the angle and length of the first, as shown in this top view.

**9.** Clip and bind down the excess leg material. Position the thread at the rearmost thread wrap securing the legs. Dub over the leg-mounting wraps to the rear base of the post. To form the optional rubber legs, mount a strand of leg material on each side of the wing post as described for the Green Drake Rubber Legs, steps 6–9, pp. 58–59, and shown here from underneath. Trim the legs to about ½ the body length.

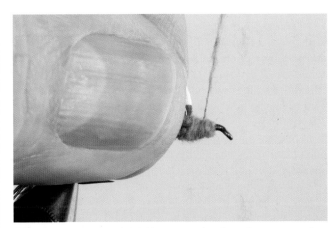

**10.** Take a wrap of dubbing over the thread wraps used to mount the rubber legs. Then draw the front pair of rubber legs rearward, and dub a thorax the same diameter as the abdomen. Position the tying thread about 4 thread wraps' distance behind the hook eye.

**11.** Wrap and tie off the hackle as described for the Parachute BWO, steps 7–10, pp. 16–17. Finish the head of the fly.

## TAN CHUGGER

| Hook: | #6-8 Mustad 94840 |
|---|---|
| **Thread:** | Yellow 6/0 |
| **Body:** | Tan or yellow Larva Lace Dry Fly Foam |
| **Cement:** | Dave's Flexament, or similar flexible cement, thinned 1-to-1 with solvent |
| **Wing:** | Tan over black Poly-Bear Fiber by Spirit River |
| **Legs:** | Brown or yellow rubber leg material |

The body of this pattern and the wiggly rubber legs float flush in the surface film like those of the naturals, but the poly down-wing with the tan yarn on top makes this fly easy to see. Over the past several seasons, it's been our top-producing pattern for both grasshoppers and stoneflies. We tie this fly in both tan and yellow versions because trout can sometimes be selective to color, or so it has seemed in our experience. If a trout rejects one color, we rest the water and then tie on the other color. This practice has accounted for enough fish that we use it after every refusal, short strike, or missed fish.

This pattern is named for the way it's fished. As the fly comes to the end of a free-floating drift and starts to drag across and downstream of you, lower the rod tip and jerk the fly underwater. Retrieve it in quick strips, or keep the rod tip low to the surface and sweep the fly through the water in quick, 2- to 3-foot bursts. We will admit this is profoundly ungrasshopperlike behavior, but the technique has worked amazingly well for us. It does, however, necessitate a durable pattern tied with materials that shed water quickly so that only one false cast is needed to dry the fly. This pattern evolved from those requirements. We generally use this technique during hopper season when fishing from a boat, but wading anglers can employ a similar method.

The procedure for tying this fly is identical to that shown for the Dirty-Yellow Chugger, pp. 197–200, and that sequence also pictures a yellow version of the pattern. For hopper imitations, however, we form only one or two body segments on the needle; the smaller body is often more effective. It isn't necessary to color the tan foam on this hopper pattern, but coat all the thread wraps used to form the body segments with cement. Instead of the Larva Lace, a strip of a similar $\frac{1}{16}$-inch (2 mm) closed-cell foam can be substituted. For information on types of foam suitable for this pattern, see p. 79.

# LOCALLY IMPORTANT TERRESTRIALS

## Moths

Moths are a diverse group of insects and are found throughout the West. Most have a terrestrial life cycle, but a number of moths are aquatic in the larval and pupal stages, though the species found in trout streams generally emerge after dark and are not considered a fishable hatch. The majority of terrestrial adult moths are nocturnal as well, but some are active during the day. And of these, two are important to anglers: the Western Spruce Budworm Moth and the Douglas-Fir Tussock Moth. Both types are found in Douglas fir, spruce, and true fir stands. Under the right conditions, moth populations can soar, becoming vast infestations that may last 3 to 7 years and seriously damage the forests. Ironically, these outbreaks can be a windfall for anglers: the unnatural density of moth populations makes the insects seasonally and locally available to trout, since many of the adults are blown or fall into the water. But even during the years with no serious infestations, moths are still well worth imitating, especially in forested headwater streams during late summer and early fall. Moths are among the largest terrestrial insects found in this type of habitat, and trout in these waters are opportunistic feeders, always alert for a meal, especially a big one. Spruce and tussock moths are not reliably present in the same way as aquatic insects; the outbreaks come and go. But for all its destructiveness, an infestation can produce some wonderful fishing—a temporary local super hatch.

**Other common names:** Miller moth, Western Spruce Moth
**Order:** Lepidoptera
**Emergence:** July to September
**Body length:** ½–1½ inches (13–38 mm)
**Hook sizes:** #12-14

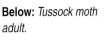
**Below:** *Tussock moth adult.*

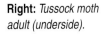
**Right:** *Tussock moth adult (underside).*

**Above:** *Brown Western Spruce Budworm Moth adult.*

**Right:** *Brown Western Spruce Budworm Moth adult (underside).*

**Left:** *Orange/brown Western Spruce Budworm Moth adult.*

**Above:** *Orange/brown Western Spruce Budworm Moth adult (underside).*

# Important Fishing Stage

**Adult** Infestations of spruce and tussock moths generally reach their peak populations from midsummer to fall, and the adults are most active in midmorning to early afternoon, when you can see them fluttering over the water like large caddis or small stoneflies. During a large outbreak, when a great many moths are available almost daily, the trout can become selective, and matching your fly to the size and color of the natural is necessary. Under these conditions, you may well find trout rising—sometimes steadily, sometimes only sporadically, but frequently enough to target them. At other times, particularly in years when moth populations are not unusually high, fishing a moth pattern is very much like grasshopper fishing—casting to the likely holding water rather than presenting the fly to individual fish. In this situation, when the trout do not routinely see large numbers of moths, a Deer Hair or Elk Hair Caddis is a reasonable approximation of the natural.

# Moth Patterns

## LAFONTAINE SPRUCE MOTH

*Originator: Gary LaFontaine*

| Hook: | #12-14 standard dry-fly |
|---|---|
| Thread: | Tan 6/0 |
| Body: | Tan or ginger mink fur dubbing mixed with guard hairs |
| Wing: | Light elk hair |
| Head: | Light elk hair |

This very simple spruce moth pattern has accounted for a lot of trout. The body rides flush against the film, like that of the natural, but the bushy, elevated wing is quite visible on the water. The spun head helps support the front of the fly and gives bulk to the silhouette, which is characteristic of both spruce and tussock moths. Mink fur is not normally sold as packaged dubbing, so this material must be clipped from a patch of hide. Don't remove the guard hairs; just mix the fur to randomize the fibers for easier dubbing. Natural mink and darker elk hair can be used to tie a darker version of the pattern. This is an easy fly to tie; if there is any trick to it, it comes in forming the head. The wing hair is held firmly atop the shank, while the thread is used to spin the hair butts around the circumference of the hook wire.

**1.** Mount the thread behind the hook eye, and wrap it to the rear of the hook shank. Dub a slightly tapered body over the rear ¾ of the hook shank. Position the thread in front of the body.

**2.** Clean and stack a bundle of elk hair as described for the Sparkle Dun, steps 1–4, pp. 27–28. With your right hand, position the bundle over the hook shank with the tips extending just beyond the bend of the hook.

**3.** Maintaining the position of the hair atop the shank, transfer the bundle to your left fingers. Trim the butt ends of the hairs even with the hook eye.

**4.** Take 2 loose wraps of thread around the hair and hook shank. Pinch the hair securely to keep it atop the shank, increase the tension on the thread, and pull the thread toward you. Then take 2 more tight thread wraps directly over the previous wraps. The wing should now be secured to the top of the hook shank, with the butt ends of the hair distributed around the shank.

**5.** Preen the hair butts rearward to stand them up almost perpendicular to the hook shank. Take 2 to 3 tight thread wraps through the hair, forward to the eye, and finish the fly around the shank behind the hook eye.

## SNOWSHOE MOTH

| | |
|---|---|
| **Hook:** | #12-14 standard dry-fly |
| **Thread:** | Tan or gray 6/0 |
| **Body:** | Dun or tan dubbing |
| **Wing:** | Dun or tan snowshoe hare's foot hair |

Although fluttering moth patterns like the LaFontaine Spruce Moth and Elk Hair Caddis can be highly effective, many moths that hit the water remain motionless, creating a wedge-shaped profile on the surface film. This flush-floating design is a more credible imitation of those insects, and it's particularly well suited to heavily fished waters, where most anglers use conventional fluttering patterns. Because of the snowshoe hare wing, this fly floats well and is quite durable. The wing and body colors can easily be changed to match the colors of the prevailing naturals. You can also add a few strands of orange or brown Antron yarn beneath the snowshoe hare to give the wings a mottled appearance when viewed from underneath. Simply mount the yarn strands, slanting rearward over the body, prior to mounting the snowshoe hare's foot in step 2. Then divide the yarn along with the wing material in step 3. When tying the fly, take care that the bundle of wing material mounted in step 2 extends more than one hook gap's distance beyond the bend to allow enough excess material for trimming.

**1.** Mount the thread behind the hook eye, and wrap it to the rear of the shank. Dub a slightly tapered body over the rear ¾ of the hook shank. Position the thread halfway between the front of the body and the hook eye.

**2.** Clip and clean a bundle of snowshoe hare's foot hair as described for the BWO Snowshoe Dun, steps 1–2, p. 13. Position the hair atop the shank so that the tips extend at least one hook gap's distance beyond the bend of the hook. Use 4 tight thread wraps to secure the bundle to the top of the hook shank.

**3.** Divide the bundle of hair into 2 equal bunches, and use 2 crisscross thread wraps to separate the wings and position them nearly perpendicular to the hook shank, as shown in this top view.

**4.** Pinch the wings along the sides of the hook shank, and wrap the thread rearward over the base of the wings, stopping one thread wrap's distance in front of the body. The wings should angle rearward at about 45 degrees, as shown in this top view.

**5.** Trim the butts of the wing hair. Dub the head, and finish the fly.

**6.** Draw both wings rearward, and trim them so that they extend one hook gap's distance beyond the rear of the body.

# Chapter Six

# Other Important Food Forms

---

## Aquatic Worms

Aquatic worms are similar in appearance to terrestrial earthworms, but they are generally shorter and more varied in color. They also resemble midge larvae but are typically larger, and unlike midge larvae, aquatic worms do not pupate and emerge. There is no "hatch." They spend their entire lives living in the detritus, silt, and debris of the streambed. Aquatic worms are widely distributed and found in nearly every freshwater environment; populations can be dense, though you seldom see these worms unless you are specifically searching for them. Trout occasionally have been observed rooting on the bottom for aquatic worms, but far more often—and of more significance to the angler—trout take worms that have been dislodged by an increase in water flow from runoff, rain, or dam releases. Wading anglers can also disturb the bottom enough to send worms into the drift, and on crowded tailwaters, trout often take up positions directly downstream of these fishermen to feed. But even under ordinary circumstances, enough aquatic worms end up in the current that trout recognize them, and worm imitations are productive.

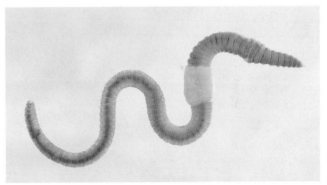

*Aquatic worm: body colors range through shades of brown, tan, black, and red.*

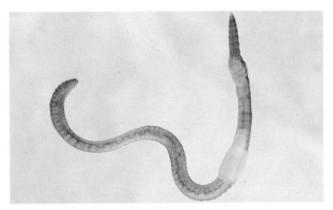

*Aquatic worm (underside): same shade as, or lighter shade of, top color; the red vein running down the body is often visible.*

| | |
|---|---|
| Other common names: | San Juan Worm, aquatic earthworm |
| Phylum: | Annelida |
| Subclass: | Oligochaeta |
| Order: | Lumbriculidae |
| Body length: | 5 inches maximum (135 mm), with 1–3 inches (25–76 mm) most common |
| Hook sizes: | Scud, #12-18 |

# Important Fishing Information

One bit of conventional wisdom among bait fishermen is that earthworms are particularly effective after a rain, the worms presumably having been washed into the water, which frankly seems a bit implausible. The more likely scenario is that aquatic worms have been disturbed from the bottom by the rain-swollen current and set loose in the drift. During and just after runoff and heavy rains are prime times to fish a worm pattern. On streams with especially rich habitat, such as some tailwaters, worm patterns are the flies of first resort. But aquatic worms are common enough almost everywhere that imitating them is a productive approach. Drifting worms are helpless in the current and are typically found close to the bottom; that's where and how an imitation should be fished— deep and drag-free. In shallower waters, a single unweighted or weighted fly may suffice, but for fishing deeper, many anglers use a worm pattern as a dropper behind a more heavily weighted fly. Colors and sizes of worms vary from stream to stream, so carrying a small range of patterns is good insurance; generally, tan, brown, and red patterns 1 to 2 inches in length will adequately approximate most naturals.

# Aquatic Worm Patterns

## SAN JUAN WORM

| Hook: | #12-18 heavy-wire scud |
| --- | --- |
| **Thread:** | Red, tan, or brown 6/0 |
| **Underbody:** | Tying thread |
| **Body:** | Red, tan, or brown Vernille or Ultra Chenille |

The San Juan Worm has two claims to fame: it is probably the single most popular worm imitation, largely because it's so effective, and it may well be the easiest pattern in all of fly tying to dress. The weight of the hook wire is often sufficient to get this fly near the bottom in shallower water, but if more weight is desired, form an underbody using wire the same color as the body material. The red-bodied version of this pattern is by far the most common, but tan and brown also work quite well.

**1.** Mount the thread behind the hook eye, and wrap a foundation to a point about $\frac{1}{3}$ of the way down the hook bend. Cut a length of body material that is 3 to 4 times the length of the hook shank. Center the material over the thread foundation, and secure it at the rear of the underbody.

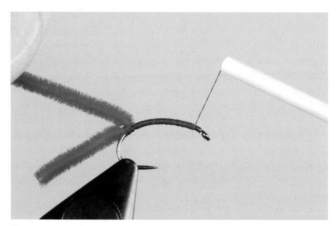

**2.** Fold the front tag of the body material rearward, and position the thread 3 to 4 thread wraps' distance behind the eye.

**3.** Fold the body material forward over the top of the shank, and secure it.

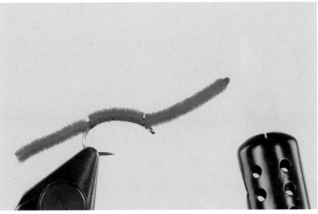

**4.** Finish the fly around the hook shank, behind the eye. Taper the front end of the body by touching the front tag of body material to the side of a flame from a lighter or match. Repeat with the rear tag of material.

## PLASTIC WORM

| Hook: | #4-12 Mustad 37160 |
|---|---|
| Thread: | Brown, tan, amber, or red 6/0 |
| Underbody: | White 3/0 or 6/0 thread, colored with black or dark brown marker |
| Body: | Medium tan, brown, amber, or red Ice Stretch Lace |

The plastic cord used on this pattern gives the fly a translucent, segmented appearance much like that of the natural, and the hook shape gives it a lifelike curvature. The bait-hook style recommended here is large for its size designation, and the heavy wire, the nonabsorbent body material, and the minimal water resistance of the design make the unweighted version of this pattern sink quickly. To get deeper still, a weighted version is often tied using an underbody of wire wrapped over the middle $2/3$ of the shank and covered completely with white thread so that the body color will show up well. The weighted version is a good point fly in a multifly rig, with a smaller nymph or worm pattern used as a dropper. A solid plastic cord works best for this pattern, giving a rounder segmentation than hollow tubing, which collapses when wrapped.

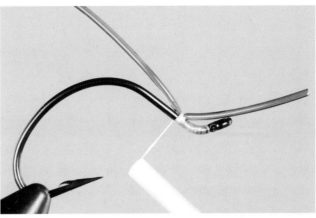

**1.** Mount the white thread behind the hook eye, and wrap it rearward one hook-eye length. Mount a length of the body material very tightly on top of the hook shank. Leave a $1/2$-inch tag of material facing forward, as shown.

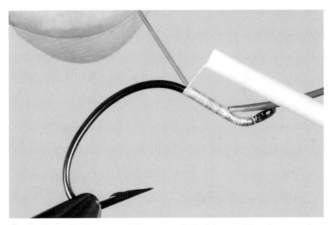

**2.** Pull the body material to stretch it thin, and begin securing it to the top of the shank.

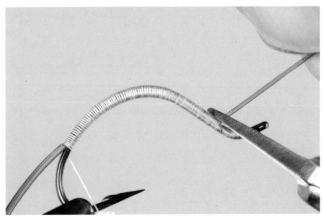

**3.** Bind the body material to the point on the shank shown here. Pull the front tag to stretch it tight, and trim it close to the thread wraps.

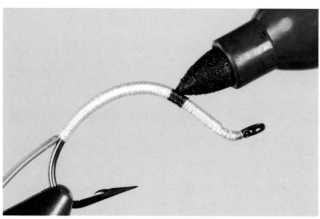

**4.** Wrap the thread forward, forming a smooth underbody. Tie off the thread behind the hook eye. With a dark brown or black marker, draw a band around the body about ⅓ of a shank length behind the hook eye.

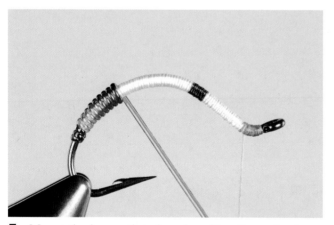

**5.** Mount the brown thread, and position it one hook-eye length behind the eye. Stretch the lace to decrease its diameter, and begin wrapping it forward in closely abutting turns. As you wrap forward, gradually relax the tension on the strand, increasing its diameter, in order to taper the rear end of the body.

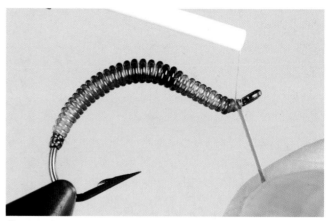

**6.** Continue wrapping the body material forward. About 4 to 8 thread wraps' distance from the hanging thread, begin to increase the tension on the strand to reduce the diameter and taper the front of the body. Secure the strand with 5 tight thread wraps. Stretch the tag end of the material, and trim it close to the thread wraps. Finish the head of the fly.

# Baitfish, Small Trout, and Minnows

Trout eat other fish, including other trout, and the bigger the trout, the more likely it is to be piscivorous. That's one of the ways it gets big. When it comes to preying on smaller fish, brown trout have the most notorious reputation, particularly for cannibalism. But in reality, rainbows, cutthroat, and especially bull trout will all consume smaller fish if they can catch them. Baitfish—the angler's informal category for minnows and other small, nongame fish—and young trout survive by sticking close to bankside cover, where they can find protection, or in the shallows near the shore, where large trout are often reluctant to pursue them, at least during the daylight hours. But small fish can be forced from safe environs by wading anglers, livestock, predators such as herons, or even the shadows of birds flying over the water. If alarmed, the fish often dart to deeper water, and then dart back to the shallows or shelter when the threat passes—a tendency that is of significance to fishermen, as imitating this panicked behavior is a highly productive way of manipulating a baitfish pattern. Along with sculpin imitations, baitfish and small trout patterns are probably the stream angler's best bet for taking a

truly large fish. Because sculpins have broad, flattened heads and rather squat bodies, they present a different profile from baitfish and small trout, so sculpin imitations are discussed in a separate section of this chapter (p. 283).

**Other common names:**  Minnows, chubs, dace
**Body length:**              2–4 inches is an effective length to imitate
**Hook sizes:**               #2-12

**Above:** *The colors of small trout (a rainbow is shown here), minnows, and other baitfish can vary a good deal, though fish that are olive-brown to brown on top and cream to light brown on the underside are common.*

*A small brown trout.*

*Dace are a common forage fish in trout streams.*

# Important Fishing Information

The small fish most consistently available to trout are generally moving either away from the bank, as they are spooked from the shallows, or toward the bank, as they return. Baitfish patterns are typically fished in the same directions. In large rivers, cast out to deeper water and retrieve the fly toward the shore. The current will sweep the fly downstream during the retrieve; when the line swings tight and straightens out below you, let the fly hang in the current for a few seconds. A trout that has followed the fly in from deeper water will often strike when it's motionless. On streams small enough to cast across, deliver the fly to the far bank and work it back to the near one. Boat anglers should cast to the bank and retrieve the fly into deeper water. Vary the retrieve speed, with and without pauses, until you find an effective combination. Unlike sculpin patterns, flies that imitate small trout and baitfish don't need to be fished close to the bottom, though it's best if they sink readily. Because these small fish are always a potential meal for trout, the imitations can fished at any time of day in virtually any type of water—riffles, runs, pools, and edge waters. But the most productive time, especially in skinny flats and pool tailouts, is in the early morning or late evening, when large trout, under the cover of low light, prowl the shallows to feed.

# Baitfish, Small Trout, and Minnow Patterns

**CLOUSER MINNOW**

*Originator: Bob Clouser*

| | |
|---|---|
| **Hook:** | #4-12 2XL–3XL ring-eye streamer |
| **Thread:** | Brown and red 3/0 or 6/0 |
| **Eyes:** | Yellow-and-black-painted lead barbell |
| **Belly hair:** | White or light tan bucktail |
| **Lateral line:** | Pearl Krystal Flash and pearl Flashabou |
| **Back hair:** | Natural brown or dyed-olive bucktail |

Originally designed as a fly for smallmouth bass, the Clouser Minnow subsequently proved its worth for virtually every species that eats baitfish. It's particularly useful on trout

streams, since the lead eyes sink the fly to fishing depth quickly, before the current can sweep it out of the sweet spot. While the pattern has some weight to it, the relatively sparse dressing and aerodynamic shape make it easier to cast on trout rods than some of the more wind-resistant baitfish patterns. The barbell eyes cause this fly to ride with the hook point up, which helps reduce snagging when fishing where small fish congregate—around cover or in shallow water. The bucktail collapses when the fly is retrieved and springs up when it's paused, giving the pattern a lifelike, pulsating action. The brown/white, brown/tan, and olive/white color combinations are good approximations of many small fish, but to better match locally prevalent baitfish species, you can replace the bucktail with stiff synthetic hair, which is available in a wider range of colors.

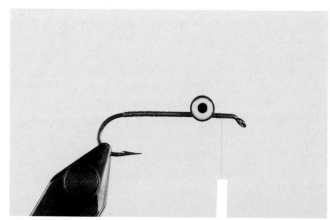

**1.** Mount the thread behind the hook eye. Wrap a foundation to the rear of the shank, and position the thread ⅓ of a shank length behind the eye. Mount the barbell eyes atop the shank, and position the thread halfway between the barbell eyes and the hook eye.

**2.** Align the tips of a sparse bundle of bucktail. Clip it from the hide, and remove any short hairs by pulling them from the base of the bundle. Mount the bucktail atop the shank with 4 to 5 tight wraps to make a wing that extends 2 shank lengths behind the mounting wraps.

**3.** Cut the bucktail butts at an angle, tapering toward the hook eye. Bind down the butt ends. Tie off the thread and clip it.

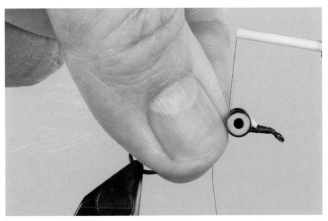

**4.** Draw the bucktail rearward along the top of the shank, between the barbell eyes. Slip the red tying thread between your left thumb and forefinger.

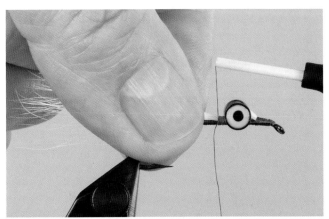

**5.** Mount the thread around the bucktail and hook shank exactly as you would mount thread on a bare hook. Take 5 to 6 tight wraps to secure the bucktail and form a narrow band of thread at the throat of the fly.

**6.** Tie off the thread and clip it.

**7.** Invert the hook in the vise. Reattach the brown thread behind the hook eye. Mount 6 to 10 strands of Krystal Flash and an equal number of Flashabou strands so that they extend one hook gap's length beyond the white bucktail. Trim and bind the excess materials.

**8.** Align, clip, and clean a sparse bundle of brown bucktail. Mount it atop the shank so that the tips are aligned with the tips of the white bucktail. Cut the bucktail butts at an angle, and secure with thread. Finish the fly. Coat the brown and red thread wraps with head cement.

## ZONKER

*Originator: Dan Byford*

**Hook:** #4-10 6XL streamer
**Thread:** Red and brown 3/0
**Underbody:** Zonker Tape
**Body:** Gold, silver, or pearl Mylar tubing
**Wing:** ⅛-inch-wide strip of black-barred gold, black-barred brown, natural brown, or natural gray rabbit
**Throat:** Grizzly hen-hackle barbs
**Head:** Five-minute epoxy
**Eyes (optional):** Yellow-and-black Adhesive Eyes

The Zonker may well have been the first trout fly to be dressed with a strip of rabbit fur for the wing, and in fact, years ago, tiers simply called any rabbit-fur strip a "Zonker strip." And though now many patterns take advantage of the undulating underwater movement of fur strips, the original fly is still an excellent one. The woven Mylar tubing creates the impression of scales and flashes like a darting minnow. The metal tape underbody produces a fish-shaped profile without adding weight, bulk, or wind resistance. The fly sinks, but not quickly, so it's a good choice for fishing shallower water. The pearl-tubing body with a gray or olive wing is a good match for small rainbow trout; a gold body and brown wing suggest a small brown trout or a baitfish. This fly can also be dressed with the metal tape underbody formed above the hook shank and the wing mounted underneath; this version rides with the hook point up to reduce snagging. When tying the fly, have a small dish of water nearby to moisten your fingertips, and use them to dampen the rabbit hair so it can be controlled more easily. The eyes are not part of the original Zonker pattern; we use them on size 4 and 6 hooks.

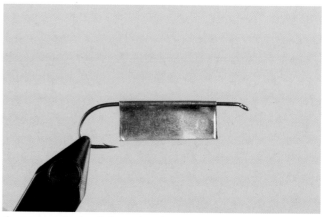

**1.** Mount the hook in a vise. Cut a strip of Zonker Tape that is ⅔ the length of the hook shank and twice the width of the hook gap. Fold the tape in half lengthwise to form a crease down the center, and then unfold it. Remove the adhesive backing. Center the crease lengthwise on the hook shank, and press the adhesive sides together.

**2.** Trim the underbody to the shape shown.

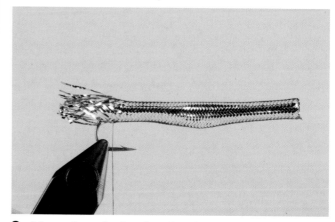

**3.** Mount the red tying thread at the rear of the underbody. Cut a piece of Mylar tubing about 1 inch longer than the hook, and remove the core. Use a dubbing needle to fray about the last ½ inch on one end. Slide the tubing over the underbody, twisting it slightly to work the frayed ends past the tying thread.

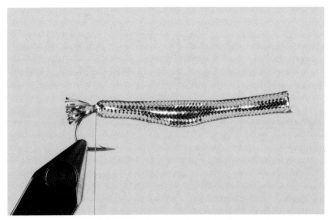

**4.** Secure the tubing to the shank with 5 to 6 tight thread wraps. If the frayed ends are long enough, trim them even with the bend of the hook.

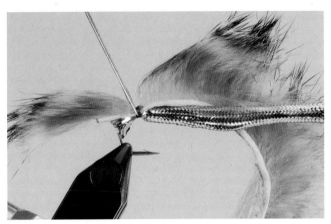

**5.** Cut a rabbit-fur strip that is about twice as long as the hook shank. The hair slants toward one end of the strip; at this end of the strip, form a gap in the fur about one hook gap's length from the end of the hide. To form the gap, dampen the hair, and slide a dubbing needle crosswise through the base of the hair at the tie-in point; draw the needle upward, and use your fingers to preen back both sides of the parted hair to expose the hide, which can then be secured to the hook shank. Mount the strip atop the shank by taking tight thread turns through the gap. Secure the thread with a few half hitches, and clip it.

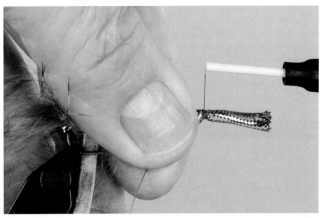

**6.** Draw the body tubing forward over the underbody. Slip the brown thread between your left thumb and forefinger, and mount the thread around the Mylar tubing directly ahead of the underbody, just as you would mount thread on a bare shank. Secure it using tight thread wraps.

**7.** Trim the thread tag and excess tubing; you might find it easier to fray the end of the tubing with a dubbing needle first, and then cut away the strands. Bind the excess tubing, and position the tying thread at the rearmost thread wrap securing the tubing.

**8.** Draw the rabbit strip forward over the top of the shank, and form a gap in the fur above the hanging thread. Secure the strip with tight thread wraps taken through the gap. Some tiers apply gel-type cyanoacrylate glue to the top of the body tubing before securing the strip for added strength.

**9.** Stretch the tag of the fur strip slightly, and trim it close to the mounting wraps. Cover the excess with thread, making a smooth, tapered foundation for mounting the hackle. Prepare and mount a hackle using the procedure shown for the Little Green Drake, steps 4–6, p. 42, but for this pattern mount the hackle with the shiny side, or front, of the feather facing upward.

**10.** Wrap the hackle forward about 4 to 5 turns as described in step 7, p. 42, and tie it off as shown in step 9.

**11.** Clip the hackle tip. Draw the barbs rearward, and take a few thread wraps over the base of the barbs to slant the hackle rearward. Finish the fly.

**12.** Coat the head of the fly with five-minute epoxy, and rotate the fly until dry.

**13.** If desired, apply adhesive-backed eyes on each side of the head. Coat the eyes and head with epoxy, and rotate until dry.

# Sculpins

Sculpins are found in both fresh and salt water, but those living in salt water or large freshwater lakes attain a greater size than those found in streams. Sculpins live among the rocks and debris on the riverbed, feeding on aquatic insect larvae, crustaceans, and even small fish. The mottled body color allows a sculpin to blend with its surroundings, but the fish is an active predator, and the trout will notice its movements in spite of the camouflaged appearance. Sculpin fly patterns—and there are a number of very good ones—probably account for more large stream trout than any other baitfish pattern. Many anglers fish a sculpin imitation in the spring, when heavy runoff dislodges sculpins from the bottom, and again in the late summer and fall, when relatively few hatches are occurring. You can, however, productively fish a sculpin pattern all season, and anglers that are interested in catching only large trout are apt to do just that. Out of reluctance to cast large, weighted flies, some fishermen regard a sculpin pattern as a backup to use when other flies fail or as a pattern of last resort. But fishing a sculpin imitation may, in fact, be the best way to catch the stream trout of a lifetime. While sculpins certainly qualify in anglers' terminology as baitfish, their distinctive body shape and habits separate them from other baitfish when it comes to tying and fishing the imitations.

| | |
|---|---|
| **Other common names:** | Freshwater sculpin, Muddler Minnow; Mottled Sculpin and Slimy Sculpin are names for two species commonly found in Western streams |
| **Family:** | Cottidae |
| **Genus:** | *Cottus* |
| **Body length:** | 6 inches (150 mm) maximum |
| **Hook sizes:** | #2-12 |

**Above:** *Sculpin (underside): cream, tan, or light brown.*

**Below:** *Sculpin (top view): sculpin are distinctly wedge-shaped when seen from above.*

Sculpin: shades of brown to olive-brown to gray body, with dark mottling.

# Important Fishing Information

The fact that sculpins prefer to hide among the streambed stones and detritus has two important consequences for anglers. First, sculpins are admirably built for this kind of existence, with pronounced pectoral fins that help stabilize them on the bottom and broad, flattened heads for living among the rocks. Both of these anatomical features are characteristic of effective sculpin imitations. And second, sculpins seldom venture off the bottom, and when they do, they rarely swim more than a few inches above the stream substrate as they move quickly from one sheltered area to another in search of food. So it's important to fish a sculpin pattern close to the streambed and with movement that mimics the natural. If the fly doesn't touch the bottom once in a while, you aren't fishing deep enough. Vary the retrieve—from fast to slow, with and without pauses—until you determine what works. Use a stout tippet, in the 1X to 3X range, and attach the fly with a loop knot so that it will move in a more lifelike manner. Sculpin patterns 2 to 3 inches long work well in most streams, but for larger rivers or those that contain the aggressively piscivorous bull trout, larger patterns are often more effective.

# Sculpin Patterns

## MARABOU MUDDLER

| | |
|---|---|
| **Hook:** | #2-12 3XL nymph hook |
| **Weight:** | Lead or nontoxic wire |
| **Thread:** | Dark brown 6/0 |
| **Tail:** | Dyed-red squirrel tail hair |
| **Body:** | Gold braid tinsel |
| **Underwing:** | Dyed-red squirrel tail hair |
| **Wing:** | Brown marabou |
| **Collar:** | Deer hair |
| **Head:** | Deer hair |

There are many versions of this pattern, which is itself an adaptation of the original Muddler Minnow, and most of them work well. The most common variations employ black or brown marabou, or a rabbit-fur Zonker strip, for the wing; the tail can be fashioned from the same material as the wing, though many tiers prefer red hair or a slip of red feather. The underwing can be tied from hair, such as calf tail, that is the same color as the wing, but we prefer a bit of red material incorporated into the body. A gold-and-brown version of this fly is pictured in the following sequence, but a silver braid tinsel body with olive marabou is also a good combination. Fished as a sculpin imitation, this fly is always weighted to help compensate for the water resistance of the large head and get the fly deep. The wide head and slender body make for a good general approximation of the shape of the natural sculpin, and the marabou gives the fly an undulating, swimming motion when fished with a strip retrieve. When tying the fly, use one of the full, fluffy marabou feathers usually labeled "marabou blood quills" or "Woolly Bugger marabou" rather than a thick-stemmed plume. The deer-hair collar can be left long, as shown in this sequence, or the hairs on the top and bottom can be trimmed flush with the back of the head to emphasize the finlike hair projecting from the sides.

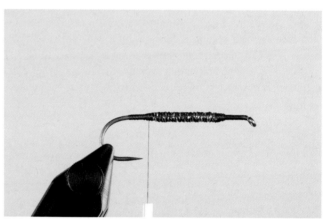

**1.** Mount and wrap the wire to create an underbody as described in the procedure for weighting the Pheasant Tail Nymph, steps 1–2, p. 21. The wire wraps should begin on the shank directly above the hook point and extend forward to a point ⅓ of a shank length behind the hook eye. Mount the thread behind the eye, and wrap rearward to the frontmost wrap of wire. Build a thread ramp in front and back of the wire wraps as shown in step 2, p. 21, and secure the wire by wrapping back and forth over it with thread. Wrap a thread foundation to the rear of the shank, and position the thread behind the last wrap of wire.

**2.** Align the tips of a sparse bundle of squirrel tail hair, and clip it from the hide. Pinch the bundle at the tip, and pull or comb out any short hairs from the base of the bundle. Trim the base of the bundle so that when it abuts the rearmost wrap of wire, the hair tips form a tail about one hook gap in length. Bind the butts of the hair tightly to the top of the shank. Position the thread behind the rearmost wrap of wire.

**3.** Position a length of braided tinsel underneath the shank so that the end of the strand abuts the rearmost wrap of wire; if you have a rotary vise, inverting the hook will simplify this operation. Mount the tinsel on the underside of the shank by wrapping rearward to the tail-mounting point. Securing the tinsel on the underside of the hook helps equalize the mounting bulk of the tinsel and hair around the shank and makes it easier to wrap the tinsel and create a smooth body. Next, build a smooth thread foundation from the rearmost thread wrap to the wire underbody. Position the tying thread at the narrow end of the front thread ramp.

**4.** Wrap the tinsel forward in slightly overlapping turns to the tying thread. Secure the tinsel and clip the excess. Wrap the thread rearward 4 to 5 wraps over the front edge of the tinsel, creating a smooth, tight thread foundation.

**5.** Align, clip, and clean another bundle of squirrel tail hairs. Position the bundle atop the shank so that the tips extend to the bend of the hook. Mount the bundle atop the shank with 5 to 6 tight thread wraps taken toward the hook eye. Clip the hair butts at an angle so that they taper down to the hook shank, and secure the butts with additional thread wraps. Position the tying thread at the rearmost mounting wrap used to secure the bundle of hair.

**6.** Select a marabou feather, and strip away the barbs at the base of the feather so that the distance between the lowermost barbs on the stem and the feather tip is about 1½ shank lengths. Mount the feather atop the shank. Cut the feather butt at an angle, as you did with the squirrel tail hair, and cover it with thread wraps to form a smooth foundation. Position the tying thread at the rearmost thread wrap used to mount the marabou.

**7.** At this point, you may find it helpful to moisten the marabou wing with your fingertip to mat the barbs together and keep them out of the way while you form the collar and head of the fly. Clean and stack a bundle of deer hair as described for the Sparkle Dun, steps 1–4, pp. 27–28. Push the center of the bundle into the hook eye so that the hair tips extend to the hook point.

**8.** Hold the hair in position, and grasp the tips with your left fingers, maintaining a uniform distribution of hair around the shank. Take 2 wraps of thread under light tension over the bundle. Maintaining your grasp on the tips, pull downward on the thread to tighten the wraps so that the hair butts flare outward. Secure the bundle with 2 to 3 additional tight wraps.

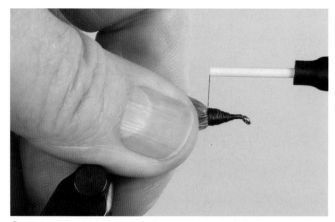

**9.** Draw the thread forward through the hair butts to the front of the bundle. Gather all the hair in your left fingers, and pull it rearward. Take 3 to 4 tight thread wraps against the base of the bundle to lock it in position.

**10.** Clip and clean another bundle of deer hair. Trim away the tips and butts so that the bundle is squared off at both ends. Push the hair into the hook eye, as you did in step 7, so that the bundle is centered at the hanging thread. Then repeat steps 8–9.

**11.** Compact the flared bundles by using your fingertips to push inward against the base of the hair from both sides.

**12.** Repeat steps 10–11, adding and compacting additional bundles until the flared hair is within 4 thread wraps' distance of the hook eye.

**13.** Finish the fly around the shank behind the eye, and clip the thread. Trim the hair, working around the hook shank, to give the head a broad, flattened shape, as shown in this top view. Take care when trimming the hair not to cut away the natural hair tips that form the collar.

## BUNNY SCULPIN

| | |
|---|---|
| **Hook:** | #2-8 TMC 300 |
| **Weight:** | Lead or nontoxic wire |
| **Thread:** | Brown 6/0 or 3/0 |
| **Eyes:** | Black-and-red Pseudo Eyes |
| **Fins:** | Brown and amber spooled Antron yarns |
| **Body:** | Tan rabbit strip and black-barred medium brown rabbit strip |

The rabbit-fur strips used on this pattern impart a sinuous, fishlike movement to the fly, while the Antron yarn represents the conspicuous pectoral fins of the natural and also suggests the wedgelike shape of the fish. The eyes may or may not enhance the appearance of this pattern to the trout; their primary purpose is to invert the hook underwater, putting the point upward, so that the fly can bounce along the bottom with fewer snags. Thus the fly is actually tied upside down, with the tan-colored belly strip on top of the hook. The body color can be changed to better match that of local sculpin populations; other popular combinations include brown and white or olive and white. When tying the fly, have a small dish of water nearby to moisten your fingertips, and use them to dampen the rabbit hair so it can be controlled more easily. Use 1/8-inch-wide rabbit strips for size 6 to 8 hooks and 1/4-inch-wide strips for size 2 to 4 hooks.

**1.** Cut the barred brown rabbit strip to 1 1/2 times the hook shank in length. The hair tips slant toward one end of the strip; trim the hide on this end to a point. Push the hook point through the center of the strip, from the hide side, 1 1/2 hook gaps' distance from the pointed end.

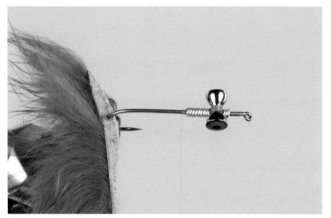

**2.** Mount the hook in the vise, and push the strip to the hook bend, as shown in this top view. Mount the thread behind the hook eye, and wrap a thread foundation over the front half of the hook shank. Position the thread ½ a hook gap's distance behind the hook eye, and mount the barbell eyes on top of the shank. If more weight is desired, wrap a wire foundation as described in the procedure for weighting the Pheasant Tail Nymph, steps 1–2, p. 21. Begin the wire foundation ⅓ of a shank length behind the hook eye. Wrap to the rear of the barbell eyes, and then take 2 to 3 wraps of wire in front of the eyes.

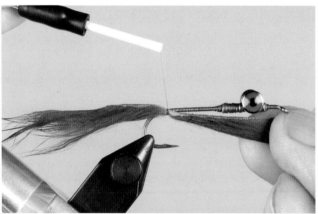

**3.** Secure the wire with thread wraps. Position the thread at the rear of the hook shank. Pull the rabbit strip up around the hook bend so that the hole in the hide is just behind the hanging thread. Form a gap in the hair at the hanging thread. To form the gap, dampen the hair, and slide a dubbing needle crosswise through the base of the hair at the tie-in point; draw the needle downward, and use your fingers to preen back both sides of the parted hair to expose the hide, which can then be secured to the hook shank. Take 3 tight thread wraps around the hide and hook shank.

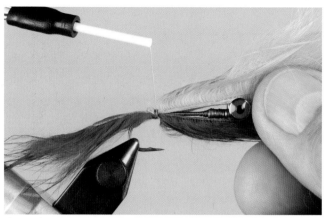

**4.** Cut a tan rabbit strip to about 1 1/2 times the length of the hook shank, and trim it to a point as explained in step 1. Use 5 tight thread wraps to mount the pointed end atop the hook shank directly over the thread-mounting wraps securing the bottom strip.

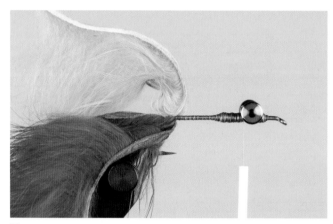

**5.** Fold both strips rearward, and wind the thread forward, stopping about 3 to 4 thread wraps' distance behind the eyes.

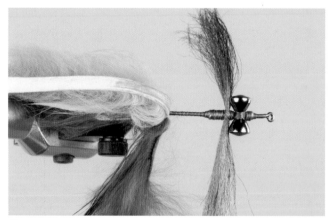

**6.** Cut two 2-inch strands of both brown and amber Antron. Combine the strands into a single bundle, and center the bundle atop the shank directly over the hanging thread. Secure the bundle with crisscross wraps, as shown in this top view.

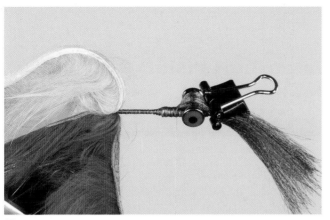

**7.** Position the thread just behind the Antron mounting wraps. Draw the Antron fibers forward, and use a small paper clamp to clip the fibers against the hook shank in front of the eyes.

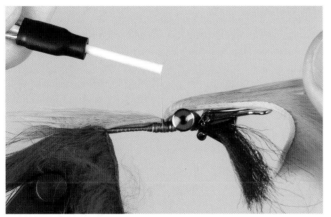

**8.** Pull the tan rabbit strip forward over the top of the hook shank. Form a gap in the hair directly above the tying thread, and secure the strip with 4 tight thread wraps.

**9.** Pull the brown strip forward. Form a gap in the hair directly below the mounting wraps securing the top strip. Secure the bottom strip with 4 tight thread wraps placed directly over the mounting wraps affixing the top strip.

**10.** Remove the paper clamp, and preen the Antron fibers outward. Wrap the thread forward to a point halfway between the barbell eyes and the hook eye. Pull the top rabbit strip over the top of the shank, between the eyes. Form a gap in the hair, and secure the strip with 4 tight thread wraps.

**11.** Trim the excess tan strip. Pull the brown strip forward on the underside of the shank, form a gap in the hair, and secure the strip with 4 tight thread wraps. Trim the excess strip, and finish the head.

**12.** Trim the Antron fibers to about 1¼ hook gaps in length, as shown in this top view.

# Fish Eggs

Fish eggs are an important food source for trout in Western streams, especially in waters containing salmon, steelhead, or whitefish, because these fish spawn between fall and late winter, when other food sources are scarce. Trout eggs are important as well, since trout will feed on their own roe. Spawning female salmon, steelhead, and trout turn on their sides and fan their tails to clean the streambed of debris and form shallow depressions, or redds, in the gravel or rocks. The female then lays her eggs in the depression while an accompanying male fertilizes them. After spawning, the female moves to the upstream edge of the redd and fans gravel back over the exposed eggs. While the eggs are being deposited and covered, some of them are swept into the current. In streams that have very large numbers of spawning fish or limited spawning areas, new redds are often formed on top of existing redds, stirring up some of the previously laid eggs and sending them drifting downstream, where trout collect them to feed on them. Whitefish don't build redds. The females hold close to the bottom over clean gravel in riffles, runs, and pools and lay their eggs; the males fertilize the eggs as they sink to the bottom and come to rest in the crevices among the rocks. Many of the eggs are swept away by the current, and trout take up positions behind spawning whitefish to feed on the eggs. Eggs of any species that have settled on the streambed can be dislodged by high water flows from rain or runoff and sent drifting downstream, making them available to trout even after spawning has ceased.

| | |
|---|---|
| **Other common name:** | Roe |
| **Spawning:** | Salmon, fall and winter; steelhead, winter; brown trout, fall through early winter; whitefish, late fall through winter; rainbow trout, spring |
| **Hook sizes:** | #6-14 |

## Important Fishing Information

The best place to fish an egg pattern is close to the bottom, downstream from spawning fish or freshly constructed redds. The clean, gravelly areas of redds are normally easy to spot; they appear as smooth, pale saucers against the darker stream substrate. Scrupulously avoid the redds when wading, as the eggs are delicate and can easily be crushed by a careless angler. Similarly, fish on the redds should be left alone. Cast just below or well to the sides of redds to avoid disturbing or snagging the spawning fish; the trout to be targeted are holding downstream of the redds, sometimes a distance of many yards. Spawning whitefish are usually not visible to the angler, but enough of their eggs are found drifting in the current that you can fish an egg pattern close to the bottom in riffles, runs, and pools and do quite well during the winter months in streams that have large whitefish populations. Most egg patterns are unweighted, like the naturals, and to get them close to the bottom, you must add weight to the leader or trail the fly behind a weighted nymph pattern. Egg imitations should be presented drag-free.

*Steelhead eggs are pale orange with a yellow tint and ¼ inch (6 mm) in diameter.*

*Trout eggs are pale orange with a yellow tint and ³/₁₆ inch (4 mm) in diameter. An eyed trout egg, like the one shown here, contains a developing embryo. Whitefish eggs (not shown) are pale pink with a yellow tint and ⅛ inch (3 mm) in diameter.*

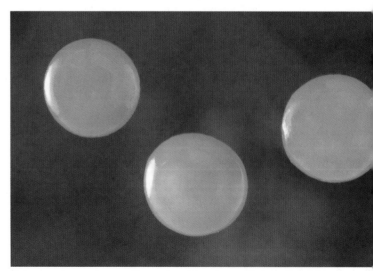

*Salmon eggs are orange with a red or yellow tint and ⅜ inch (9 mm) in diameter.*

# Fish Egg Patterns

## HOT GLUE EGG

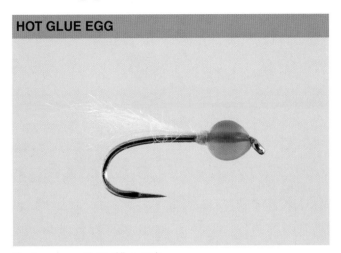

| | |
|---|---|
| **Hook:** | #8-16 1XL nymph |
| **Thread:** | White 8/0 |
| **Tail:** | Milky White Egg Veil |
| **Egg:** | Champagne or salmon hot glue sticks |

Forming this egg pattern may be less like conventional fly tying than it is like model building, and as a consequence, many tiers and anglers do not consider it a legitimate fly. Nor do some government agencies that establish fishing regulations. In some states, curious as it may seem, simply adding a tail to this pattern makes the difference between a fly and a lure and thus may qualify it for use in fly-only waters. To those who reject this kind of "plastic fly" from start, a tail does not legitimize it. Part of the resistance to this fly probably comes from its superficial resemblance to the plastic beads that are pegged to the leader above a bare hook to fish anadromous species in some areas—an approach that frequently results in snagged fish. But in that respect, this fly is innocent, no more a snagging pattern than an ordinary nymph is. Moreover, it is easy and inexpensive to make, it sinks readily, and it catches fish. Adding or omitting the "tail"—actually an imitation of a bit of skein tissue adhering to the egg—may be more a matter of local legality than of effectiveness. You can get creative with the hot-glue pattern or by layering the colors, adding more eggs to form a cluster, or coating a glass or metal bead to make a two-toned or weighted egg.

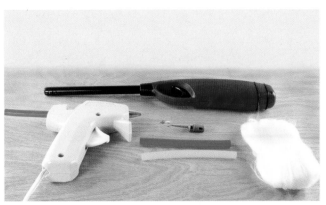

**1.** To make this pattern, besides the hook and tying thread, you need a ⁵⁄₁₆-inch mini glue stick in your desired color, a mini hot glue gun, Milky White Egg Veil (you can substitute white marabou), a long-necked butane lighter to correct the shape of the egg if necessary, clip-type hackle pliers to grip and rotate the hook, and a glass of cold water in which to cool the glue after the egg is formed.

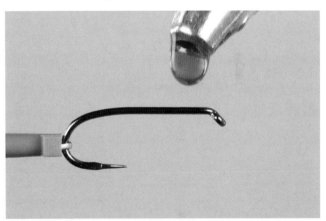

**2.** Follow the instructions that come with the glue gun for loading the glue stick and dispensing the glue. Mount the hook in the pliers, and squeeze out a drop of glue at the tip of the gun. For a small egg, the drop should be about the size of the egg to be formed. If you're making a large egg, it's better to start with a small amount of glue and apply additional layers until the egg is the desired size.

**3.** Touch the glue to the top of the hook shank about ¼ of a shank length behind the eye. Forming the egg farther back may obstruct the hook gap and make it difficult to hook fish.

**4.** While keeping the tip of the gun in contact with the glue, draw the glue to the bottom of the hook shank and up the other side.

**5.** When you reach the top of the hook shank, pull the gun away from the glue. The glue should now encircle the hook shank. Rotate the hook to distribute the glue while it is still warm, until the desired shape is formed. If more glue is needed, squeeze out another drop, apply it to the top of the egg, and repeat steps 4–5.

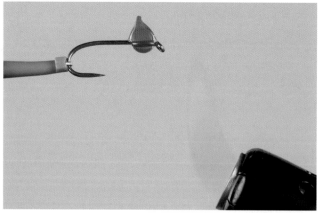

**6.** If the egg is poorly formed and the glue has cooled so that it's no longer pliable, use the butane lighter to heat the glue back to a workable viscosity. Do not put the glue directly into the flame or it will burn; hold the egg about 2 inches above the flame, and rotate the hook so that the glue is evenly heated.

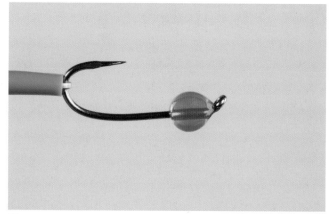

**7.** As soon at the glue becomes pliable again, remove the lighter and continue rotating the hook until the desired shape is formed.

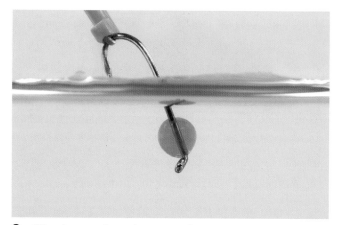

**8.** Dip the egg into the water for a moment to cool and harden the glue.

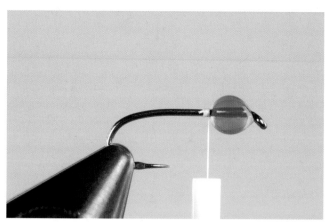

**9.** To add a tail, place the hook in the vise, and mount the tying thread directly behind the egg.

**10.** Pull a small amount of the tail material from the skein, and center the strand above the tying thread. Mount it with 3 tight thread wraps.

**11.** Fold the forward-facing fibers to the rear, and secure them with 3 thread wraps placed directly over the original mounting wraps.

**12.** Finish the fly with a few half hitches behind the egg, and clip the thread. With your left fingers, firmly pinch the tail material at the point where it's aligned with the hook bend. With your right fingers, pinch the tail material just behind the bead. Use your left thumbnail to tear the material and form a tapered end that reaches to the bend of the hook. On smaller flies, it's simpler to tear the tail to the desired length before mounting it. After completing the tail, apply a drop of head cement to the thread wraps.

## GLO BUG

| | |
|---|---|
| **Hook:** | #4-12 scud |
| **Thread:** | Red 6/0 |
| **Body:** | Orange Glo Bugs or Egg Yarn |

The Glo Bug is certainly among the most popular egg patterns for Western waters. It's easy to tie, and though producing nicely spherical eggs takes a bit of practice, even imperfectly formed ones will take fish. Some anglers speculate that fish hold on to this pattern longer than they do an egg imitation fashioned of glue or a plastic bead because the yarn is soft in texture. One of the keys to tying this pattern is using relatively few, but very tight, thread wraps. Egg yarns are thick and slightly abrasive, and if you find that the thread breaks often, go to a heavier thread size or even to gel-spun poly thread, which is extremely strong and can withstand very high wrapping tension. The other key is to trim the yarn with a single smooth cut, which is best accomplished with sharp, heavy-bladed scissors. The amount of yarn needed varies with hook size. For larger flies, use three lengths of yarn stacked together and mounted as a single bundle. For smaller hooks, like the size 12 shown in the following sequence, only a single length of yarn is necessary. Some anglers tie this fly by placing a yarn strand of a contrasting color, usually red, on top of the one or two other yarn strands to make a bundle; when the egg is clipped to shape, the contrasting strand appears as a round dot, suggesting an eyed egg that contains a developing embryo. This pattern sinks rather slowly, so add weight to the leader as needed to get the fly near the bottom.

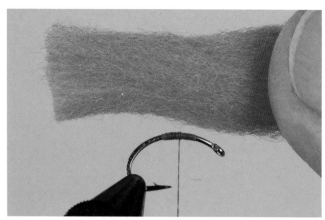

**1.** Wrap a thread foundation over the middle ⅓ of the hook shank, and position the tying thread in the middle of the foundation. Clip a 1½-inch length of yarn. (For larger hooks, clip additional lengths of yarn, and gather them side by side to make a single bundle.) Position the yarn atop the shank, centered over the hanging thread.

**2.** Mount the yarn atop the shank using 4 very tight thread wraps around the center of the bundle. Do not let the yarn slip around the hook shank.

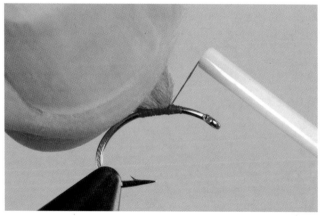

**3.** Pull all the yarn rearward, and take 2 tight thread wraps against the front base of the mounted bundle.

**4.** Draw all the yarn forward, and take 2 tight thread wraps against the rear base of the bundle.

**5.** Finish the fly around the shank ahead of the mounted yarn, and clip the thread. Smooth the bundle of yarn upward, and hold it vertically. Position the scissors horizontally to trim the yarn to a height of about ½ the hook gap.

**6.** Trim the yarn. Use your thumb and forefinger to fluff the yarn down around the sides and underside of the shank to form a sphere. If necessary, neaten up the fly by trimming down any high spots or stray fibers.

# Leeches

Leeches are widely distributed in freshwater streams, living among the rocks and debris on the streambed, and like aquatic worms, they are more prevalent than anglers often realize. The body of a leech is quite elastic and can extend to the familiar ribbonlike shape or contract into a ball; should you pick up a rock from the stream and find a soft, fleshy hemisphere attached to the underside, it's probably a leech. The majority of species are not the hematophagous, or bloodsucking, type, but feed by scavenging or preying on aquatic worms, insects, crustaceans, and mollusks. Suckers located at both ends of their bodies allow them to move inchwormlike across the bottom, and when they detach themselves, leeches are fairly agile swimmers, moving with a sinuous, up-and-down motion. Leeches are sensitive to light and generally stay hidden until the sun leaves the water. But in most streams, especially those with good populations, leeches can be found drifting or swimming in the current often enough during daylight hours that trout become accustomed to seeing and feeding on them.

| | |
|---|---|
| **Other common name:** | Bloodsucker |
| **Phylum:** | Annelida |
| **Class:** | Hirudinea |
| **Body length:** | 6 inches (15cm), with 2–4 inches (50–100 mm) most common |
| **Hook sizes:** | #2-12, 3XL |

# Important Fishing Information

Early mornings and evenings, when the naturals are more prone to leave the shelter of the streambed, are arguably the best times to fish a leech pattern. But the daylight hours can offer good fishing as well. Because of the strong suckers at the ends of their bodies, leeches are not normally swept away by the current, but they do appear in the drift when they are searching for food or moving from place to place. They are more likely to be encountered in streams or sections of streams where the current is not strong, and when moving about, they remain close to the bottom, so leech patterns are usually fished deep in waters with slow to moderate flows. To imitate the undulating, swimming motion of the natural, use a pattern with a weighted head and sinuous, flowing tail. Fish the fly with a strip retrieve or twitches of the rod tip; with each pause during the retrieve, the weighted head will dip, and then it will rise again with the next strip or twitch. Vary the speed of the retrieve, the length of each strip or twitch, and the duration of the pause until you find a combination that brings strikes.

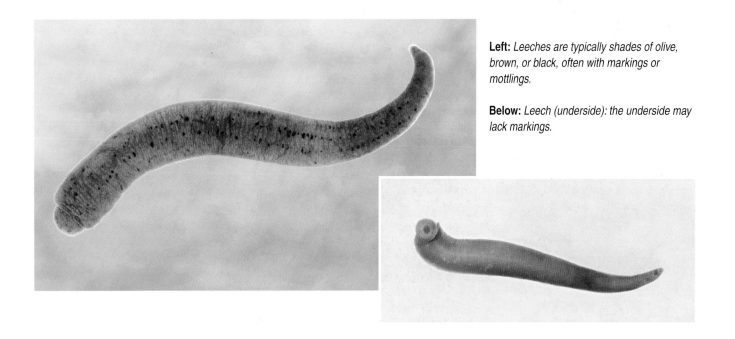

**Left:** *Leeches are typically shades of olive, brown, or black, often with markings or mottlings.*

**Below:** *Leech (underside): the underside may lack markings.*

# Leech Patterns

## BEADHEAD WOOLLY BUGGER

*Originator: Russell Blessing*

| | |
|---|---|
| **Hook:** | #2-14 3XL–4XL streamer |
| **Head:** | Black, gold, or red metal bead |
| **Weight:** | Lead or nontoxic wire |
| **Thread:** | Black or brown 6/0 |
| **Tail:** | Black or brown marabou |
| **Rib:** | Fine gold wire |
| **Hackle:** | Black or brown wet-fly or streamer hackle |
| **Body:** | Black or brown chenille |

The addition of a metal bead on this widely used pattern provides the off-center weight necessary to give the fly a swimming, jigging motion that can't really be achieved with an underbody of wire, which tends to deaden the action of the fly. Marabou feathers usually packaged as "blood feathers" or "Woolly Bugger marabou" are best for this fly, since the long, fuzzy barbs and fine stems maximize the mobility of the material underwater. As full and fluffy as marabou looks when it's dry, in the water the barbs collapse dramatically, and an insufficient amount of material will make an anemic, threadlike tail. On larger hooks, two or more feathers may be needed to dress the tail. Very soft, long-barbed hackle also promotes movement in the fly. Schlappen hackle is excellent for this pattern, but it often has very long barbs that are proportioned for larger hooks. A better bet for smaller hooks is strung wet-fly or streamer hackle, or very webby saddle hackle. Leeches scavenge fish roe, and when trout, salmon, or steelhead are spawning, from fall to spring, a fly with a red bead at the head can be used to imitate a leech that is feeding on a fish egg.

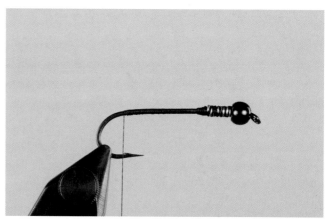

**1.** Mount the bead and 6 to 8 turns of wire around the shank, and form a tapered thread ramp at the rear of the wire as described for the Beadhead Prince, steps 1–3, p. 141. Wrap the thread to the rear of the shank.

**2.** Mount a marabou feather atop the shank to make a tail about one hook shank in length. Secure the butt end of the feather to the top of the shank by wrapping forward. Trim the feather at the rearmost wrap of wire, and bind down the excess. Position the thread at the rearmost wrap securing the tail.

**3.** Clip a length of chenille. Use your thumbnail to scrape away the fibers from one end and expose about ¼ inch of the thread core.

**4.** Mount the chenille atop the shank by binding down the thread core. Return the thread to the rearmost mounting wrap, and secure a length of gold wire atop the shank. Position the tying thread behind the bead.

**5.** Wrap the chenille forward in close, tight turns. Secure the strand behind the bead, and clip the excess.

**6.** Prepare and mount a hackle feather as shown in the instructions for the Small Green Drake, steps 4–6, p. 42, but for this fly, mount the feather with the shiny, front side facing the hook eye. Trim the excess feather stem, and position the thread behind the bead.

**7.** Using your fingers or a pair of hackle pliers, take 2 wraps of the hackle feather directly behind the bead, with the second wrap directly abutting the rear of the first one. Then wrap the hackle feather rearward to the tail-mounting point in an open spiral of 6 to 8 turns. As you wrap rearward, keep the feather perpendicular to the hook shank so that the barbs splay evenly outward.

**8.** Hold the hackle or hackle pliers above the hook shank in your left fingers. With your right fingers, take 2 turns of the ribbing wire over the tip of the feather to secure it.

**9.** Spiral the ribbing wire forward with same spacing as the hackle wraps. As you wrap the wire, wiggle it slightly to push the hackle barbs aside, and seat the wire against the feather stem without binding down the barbs. When you reach the bead head, secure the wire, and trim the excess. Clip away the feather tip. Finish the fly behind the bead.

## CONEHEAD BUNNY LEECH

| Hook: | #4-10 2XL–3XL streamer |
| Head: | Black, copper, or gold conehead |
| Weight: | Lead or nontoxic wire |
| Thread: | Dark brown 6/0 |
| Tail: | Dark brown, brown, or black rabbit-fur strip |
| Body: | Dark brown, brown, or black rabbit-fur strip |

Rabbit strips are an ideal material for leech patterns; the soft hair and flexible hide are extremely mobile underwater and reproduce the swimming behavior of the natural. Some leech designs maximize this mobility by using a long rabbit-fur strip lashed to the hook. But long, trailing strips can encourage short strikes and fouling on the cast, so we prefer this wrapped-body design with a fairly short tail. It still has good action in the water and produces reliable hookups. A black metal conehead best represents a natural leech color, but the shinier brass and copper types, though less realistic in appearance, work quite well, perhaps adding the kind of flash that appeals to trout in a bead-head nymph. Coneheads sink the fly well and keep it down so that it can be fished with movement and still remain near the bottom. To maximize the action in this fly, attach it to the leader with a loop knot, which allows it to move more freely.

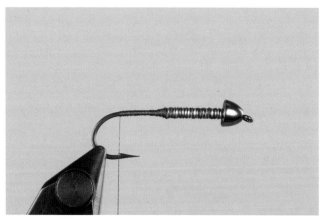

**1.** Mount the conehead and form a foundation of wire using the instructions for the Beadhead Prince Nymph, steps 1–3, p. 141. When the wire is secure, wrap a tight thread foundation to the rear of the shank.

**2.** The hair on the rabbit-fur strip slants away from one end; trim the hide on this end to a point.

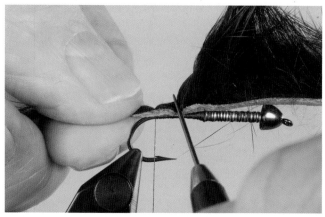

**3.** Position the strip atop the shank so that the pointed end forms a tail about one hook gap in length. Use a dubbing needle to part the fur crosswise directly above the tying thread. Dampening the hair makes the fibers easier to part and keeps the gap open.

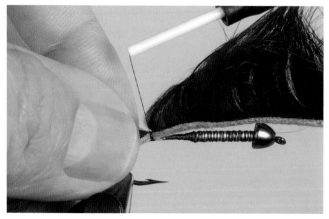

**4.** Mount the strip atop the shank with several thread wraps over the exposed hide. Make these thread wraps very tight, or the torque that comes from wrapping the body strand will cause the tail to slip around the shank.

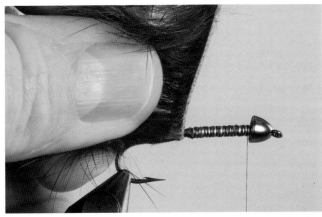

**5.** Lift the front of the strip, and position the thread behind the conehead. Wrap the strip forward in edge-to-edge wraps with no gaps. As you make each wrap, use your left fingers to smooth the hair from the previous wrap rearward so that it isn't trapped beneath the new wrap.

**7.** Secure the strip with tight thread wraps taken over the exposed hide. Trim the excess strip, bind the butt end with tight thread wraps, and finish the fly behind the conehead.

**6.** Take the last wrap directly behind the conehead. Use a dubbing needle to part the fur directly behind the head.

# Scuds

Scuds in streams live in the shallow, debris-strewn bottoms of riffles, runs, and pools with slow to moderate flows, and they are particularly abundant where aquatic vegetation grows in thick, matlike tangles. These crustaceans require dissolved calcium in the water to build their exoskeletons; they grow by molting these shells, and thus require a constant supply of calcium to form new ones. For this reason, dense populations of scuds can be found in nutrient-rich, alkaline waters such as many spring creeks and tailwaters. Female scuds carry their orange-colored eggs and newly hatched young in brood sacs on their undersides until the hatchlings have developed enough to be released. Scuds have multiple broods throughout the summer, and because successive generations are present in the water, scuds of various sizes are routinely available to trout. Like many crustaceans, scuds are light sensitive and normally stay hidden during the day, becoming active after the sun leaves the water. They swim tilted on their sides, typically moving 1 to 6 inches and then pausing before advancing again. When scuds die, their bodies turn orange and often end up drifting in the current; fish take them, and orange-colored flies are sometimes used to imitate them.

| | |
|---|---|
| **Other common name:** | Sideswimmer |
| **Order:** | Amphipoda |
| **Body length:** | ¼–¾ inch (5–20 mm), with ½ inch (12 mm) a good size to imitate |
| **Hook sizes:** | #6-16 |

# Important Fishing Information

Scuds swim by stretching their bodies almost straight and propelling themselves forward a short distance with their legs, then resting motionless for a moment, during which time they sink a bit, partially curling their bodies. For drag-free nymph fishing, a pattern tied on a scud hook better imitates this semi-curled position. When fishing a fly with a retrieve, a pattern tied on a straight hook more closely represents the stretched body of a swimming scud. A popular way to fish a swimming scud pattern is to trail it a foot behind a streamer or leech pattern, using a moderate to slow retrieve with frequent pauses at short intervals; scuds do not swim far before resting a moment. An orange scud pattern imitating a dead scud can be tied on a curved or straight hook, but it should always be fished drag-free, since dead scuds don't swim. Whether fished on a dead drift or by imparting movement to the fly, a scud pattern is generally most effective when fished close to the bottom.

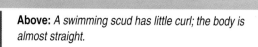

**Above:** *A swimming scud has little curl; the body is almost straight.*

**Top left:** *A scud at rest is slightly curled. Body colors range from shades of olive to tan to gray; a female carrying eggs has an orange spot in the lower-middle region of the body.*

**Left:** *A dead scud turns a shade of orange.*

# Scud Patterns

**SPARKLE SCUD**

| Hook: | #8-16 scud or 1XL nymph |
|---|---|
| Weight: | Lead or nontoxic wire |
| Thread: | Light brown 6/0 or 8/0 |
| Tail: | Light brown mottled hen-hackle barbs |
| Shellback: | Tan ⅛-inch Scud Back |
| Rib: | Fine copper wire |
| Body: | Olive-brown, olive, tan, gray, or rusty orange Ice Dub |
| Antennae: | Light brown mottled hen-hackle barbs |

The back of a natural scud is smooth and distinctly segmented, and this pattern uses a glossy shellback material that is ribbed with copper wire to suggest the segmentations. Sparkle-type dubbing adds a bit of flash to the pattern, though we probably use it for reasons closer to superstitious personal preferences than any real logic. Ordinary dubbings produce good results as well. A band of orange dubbing can be used for the middle of the body to imitate an egg-laden female, and some anglers find this variation to be important in their fishing. This fly can be tied on a straight or curved hook to imitate a swimming or resting scud, but in either case, tease out a good bit of dubbing on the underside. In addition to imitating legs, the dubbing creates a body silhouette that is broad in profile but relatively narrow when viewed head-on, mimicking the shape of the natural.

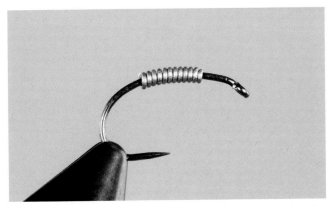

**1.** Place a hook in the vise, and wrap lead or nontoxic wire over the middle half of the hook shank as described for the Pheasant Tail Nymph, step 1, p. 21.

**2.** Mount the thread behind the hook eye. Build a tapered thread underbody, as shown, from the rear of the hook eye to the tail-mounting position. Position the thread at the rear of the underbody. Align the tips of 6 to 10 hen-feather barbs, trim them from the feather stem, and mount them to form tails about one hook gap in length.

**3.** Trim the excess tailing material. Clip one end of a length of Scud Back to a point. Mount the material by the point atop the shank directly over the tail-mounting wraps.

**4.** Place a small amount of dubbing on the tying thread, and take 2 to 4 wraps over the thread wraps used to mount the tail and shellback. Mount a length of copper wire close to the bottom of the hook shank on the near side, directly in front of the dubbing.

**5.** Dub the rest of the body, stopping 8 to 10 thread wraps' distance behind the hook eye. Position the thread at the front edge of the body.

**6.** Pull the shellback over the top of the body. Secure it in front of the body with 5 tight thread wraps.

**7.** Trim the excess shellback material, and position the thread midway between the front of the body and the hook eye. Spiral the wire forward in 6 to 8 wraps, and secure it in front of the body.

**8.** Trim the excess wire, and position the thread about 5 thread wraps' distance behind the eye. Align the tips of 6 to 10 hen-feather barbs, clip them from the stem, and mount them atop the shank to form antennae about one hook gap in length. Trim the excess feather, and cover the butt ends with thread. Lift the antennae, and form a neat head.

**9.** Finish the fly around the shank behind the eye, and clip the thread. Use a dubbing needle to tease out strands of dubbing from the underside of the body. Trim any fibers that are excessively long.

## BRUSH SCUD

| | |
|---|---|
| **Hook:** | #8-16 scud or 1XL nymph |
| **Thread:** | Light brown 6/0 or 8/0 |
| **Tail:** | Medium brown or medium dun hen-hackle barbs |
| **Dubbing brush core:** | Fine gold wire |
| **Body:** | Tan Ice Dub |
| **Antennae:** | Medium brown or medium dun hen-hackle barbs |
| **Shellback:** | Five-minute epoxy |

Wire-core dubbing brushes form bodies that are shaggier and more bristle-like than those made with ordinary twisted-dubbing methods, and they are stronger than thread-core dubbing loops. Choose a medium-textured, relatively long-fibered dubbing. Extremely fine dubbings, such as many of the poly types, don't produce the rough, leggy appearance that is desirable on a scud pattern. Instead of the Ice Dub used here, which is a combination of natural and synthetic materials, you can substitute a synthetic material such as Antron, whose crinkly fibers make realistic-looking legs. This pattern can also be weighted with a wire underbody and tied in olive, olive-brown, gray, or rusty orange.

**1.** Mount the thread behind the hook eye, and wrap to the midpoint of the shank. Cut a 6-inch length of gold wire, and align the ends. Mount the ends atop the shank, wrapping the thread forward to the hook eye. Insert the middle fingers of your left hand into the loop that's been formed in the wire to hold the wire strands apart. Apply dubbing wax to both wire strands. Hang a dubbing hook between your left fingers, at the base of the wire loop.

**2.** Hold a ball of dubbing in your left fingertips. With your right fingers, pull a pinch of dubbing, and center it on the lower waxed wire about ½ inch below the hook shank. It will stick to the wax.

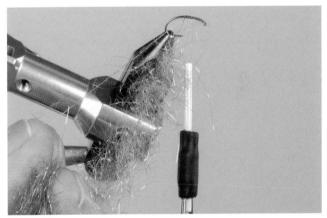

**3.** Add more pinches of dubbing, working down the wire toward the base of the loop, until the dubbing covers about 2 inches of the wire.

**4.** Remove your fingers from the loop so that the dubbing is trapped between the wires. Spin the dubbing hook 8 to 10 times (clockwise when viewed from above). Don't overtwist the wire or it will break, though if the wire does break, the dubbing brush is still usable.

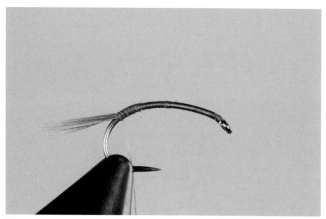

**5.** Unwrap the tying the thread to free the dubbing brush from the hook shank. It will not untwist. Leave the dubbing hook in the wire loop. Wrap the thread to the rear of the hook. Align the tips of 6 to 10 hen-feather barbs, trim them from the feather stem, and mount them to form tails about one hook gap in length. Trim the excess, and position the thread at the rearmost mounting wrap.

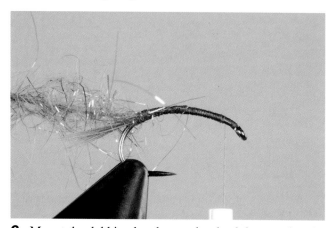

**6.** Mount the dubbing brush atop the shank by securing the wire directly ahead of the spun dubbing. Bind the tag of the wire core atop the hook to the midpoint of the shank, and clip the excess. Position the thread 6 to 8 thread wraps' distance behind the eye.

**7.** Wrap the dubbing brush forward, using the handle of the dubbing hook. On each wrap, preen the fibers from the previous wrap rearward so they won't become trapped under the new wrap of the dubbing brush.

**8.** When you reach the tying thread, tie off the dubbing brush, and trim the excess by wiggling the wire core until it breaks. Align the tips of 6 to 10 hen-feather barbs, trim them from the feather stem, and mount them to form antennae about one hook gap in length. Trim the excess, and bind the butts with thread. Lift the antennae, and finish the fly around the shank behind the hook eye. Use your fingers to fluff out the dubbing all around the body so that it stands out radially.

**9.** Trim away the dubbing on the top and sides of the body, clipping very close to the hook shank. Trim away any fibers that obscure the tails or antennae, taking care not to cut the feather barbs forming these components. Clip the dubbing fibers beneath the shank about even with the hook point

**10.** Once the fly is trimmed, use a toothpick to apply a band of five-minute epoxy along the top of the body and just slightly down the sides.

# Sow Bugs

Sow bugs resemble scuds in a number of important respects. Both are crustaceans that inhabit detritus on the streambed and tangles of aquatic vegetation in shallower streams or sections of streams with slow to moderate currents, and in fact, they often live side by side. They require calcium-rich water in order to build their exoskeletons, they grow by molting, and they can be most abundant in spring creeks and tailwaters. They have multiple broods per season and can achieve high population densities. Unlike scuds, however, sow bugs are poor swimmers and crawl about on the bottom to forage and feed. Though they normally attempt to stay hidden in the weeds and debris, many are swept into the current by surges of water from rain, runoff, and dam releases, or are dislodged by wading anglers. They drift downstream until they settle to the bottom, during which time they are easily captured by trout. In streams where they occur in large numbers, sow bugs are a significant food source.

| | |
|---|---|
| Other common name: | Cress bug |
| Order: | Isopoda |
| Body length: | ¼–¾ inch (5–20 mm), with ⅜ inch (8 mm) a good size to imitate |
| Hook sizes: | #6-16 |

*Sow bug (top view): shades of brown, gray, and yellow.*

*Sow bug (side view): note the flattened body, unlike the broad, shrimplike profile of a scud.*

# Important Fishing Information

Scuds and sow bugs are related and resemble one another superficially, but they are shaped differently. Whereas a scud is shrimplike in appearance, an aquatic sow bug looks much like the terrestrial variety, which is also known as a wood louse, pill bug, or "roly-poly." Sow bugs are best imitated with a fly that is relatively broad and flat. Because the naturals are helpless in the current, sow bug patterns should be fished dead-drift, close to the bottom; the margins of weedy areas and the waters just downstream are prime territory. In the shallows, use an unweighted imitation tied on a heavy-wire hook suspended just above the bottom by a dry fly or small indicator. In deeper water, use a weighted pattern, add weight to the leader, or fish a sow bug pattern as a dropper behind a larger weighted nymph.

# Sow Bug Patterns

**SOW BUG**

| | |
|---|---|
| Hook: | #14-16 scud |
| Weight: | Lead or nontoxic wire |
| Thread: | Tan 8/0 |
| Shellback: | Gray ⅛-inch Scud Back |
| Tails: | Dyed-gray or natural goose biots |
| Rib: | Tan 6/0 thread |
| Body: | Wapsi Sow-Scud Dubbing, light or dark sow bug color |

This sow bug imitation is simple to tie and fishes well. It has enough weight to penetrate the surface film easily and sink, but it is not so heavy that it can't be used as a dropper behind a dry fly or small indicator. Materials wrapped around a hook shank are radially symmetrical, and producing a broad, flat body profile takes a bit of inventiveness. This pattern solves that problem by using a dubbed body that is flattened on top by the shellback material and clipped flat underneath. The trick is to use a generously dubbed thread so that enough material remains after trimming to suggest the shape of a natural sow bug. You can also tie this pattern unweighted on a size 18 hook with 8/0 thread for the rib.

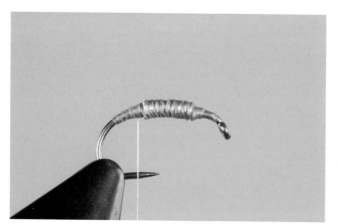

**1.** Form a foundation of wire over the middle half of the hook shank, secure the wire with thread wraps, and build a tapered thread ramp at each end, as described for the Pheasant Tail Nymph, steps 1–2, p. 21. Position the thread behind the rearmost wrap of wire.

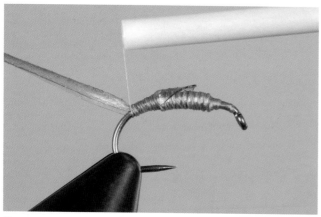

**2.** Cut a length of Scud Back, and trim one end to a point. Position the strip atop the hook shank so that the pointed end extends to the midpoint of the wire foundation. Mount the strip by wrapping rearward to the position shown, keeping the strip centered along the hook shank.

**3.** Apply a small amount of dubbing to the thread, and form a small dubbing ball halfway to the rear of the wire foundation.

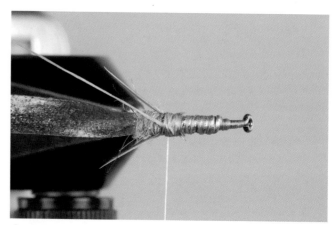

**4.** Mount a goose biot on each side of the hook shank, directly ahead of the dubbing ball, so that the tips extend rearward a distance of one hook gap, as shown in this top view. Trim the biot butts. Mount a length of 6/0 thread on top of the hook shank over the biots, and position the bobbin thread at the rearmost wrap securing the biots.

**5.** Use a heavily dubbed thread to form a body that ends just in front of the wire foundation.

**6.** With a dubbing brush or needle, pick out the dubbing on top of the body, and preen it evenly to the sides, as shown in this front view.

**7.** Pull the Scud Back over the top of the body, and secure it with 5 tight thread wraps. On smaller hooks, stretch the Scud Back as you pull it to make the material narrower and better proportioned to the body size.

**8.** Spiral the ribbing thread forward in 4 to 6 turns, and tie it off in front of the Scud Back.

**9.** Trim the excess materials, and finish the fly. With a dubbing needle, carefully pick out the dubbing along the edges of the Scud Back so that the fibers spread outward on either side; trim these fibers to about one hook gap in length. Then clip away any fibers on the underside of the shank, as shown in this front view.

## RAY CHARLES

*Originator: Bob Krumm*

| | |
|---|---|
| **Hook:** | #14-18 scud |
| **Thread:** | Red 6/0 |
| **Shellback:** | Medium or large pearl tinsel |
| **Body:** | Gray ostrich herl |

This sow bug imitation ties up quickly and easily. It was originated for fishing Montana's Bighorn River, but it has since proven its worth on a great many waters where sow bugs are found. Oddly, we've never seen this pattern tied in a weighted version, though we usually fish it with a bit of weight on the leader behind a strike indicator. Some tiers rib this fly, usually with 2-pound-test monofilament, to increase the durability—a necessity if the ostrich strands are simply wrapped around the hook shank. But twisting the herl strands with a thread core, as shown in the following sequence, produces a tough body from a fragile material. Choose herl strands with fibers that are about ½ the hook gap in length for the body. The shellback can be formed from a narrower or wider strand of tinsel; we're using a wider one here to flatten out the body profile. Other versions of this fly are tied using olive, tan, and pink herl.

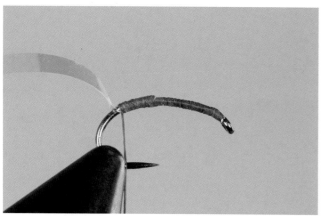

**1.** Mount the thread behind the eye, and wrap a thread foundation to the rear of the shank. Mount a length of tinsel atop the shank, and return the thread to the rearmost mounting wrap.

**2.** Align the tips of 2 to 4 ostrich herls, and gather them into a bundle along with a 6-inch scrap of 6/0 tying thread in gray or a color to match the herl. Clip the bundle ½ inch from the tip. Mount the bundle atop the shank, bind down the excess, and position the thread about 8 thread wraps' distance behind the hook eye.

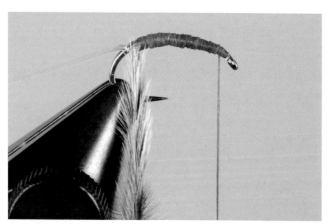

**3.** Draw the herls and scrap thread downward as a bundle, and gently twist the bundle 4 or 5 times to form a chenille-like strand; don't be too aggressive, or the fragile herls will break.

**4.** Wrap the bundle forward to the tying thread, pausing to retwist it as necessary to maintain a bushy strand. Finish the last wrap of the body with the herl strand held vertically above the shank. Secure the strand with 2 to 3 tight wraps of thread.

**5.** Clip the excess herl strand, and secure the butts with thread wraps. Position the thread at the front of the body. Bring the tinsel strand forward over the top of the body, and secure it with 3 to 4 tight thread wraps, as shown in this top view.

**6.** Clip the excess tinsel, form a thread head, and finish the fly. Here's a front view of the pattern.